Name: _____

11+

Non-Verbal Reasoning

Workbook
Age 9 – 11

Alison Primrose

GALORE PARK

AN HACHETTE UK COMPANY

Orders: **Teachers** please contact Hachette UK Distribution, Hely Hutchinson Centre, Milton Road, Didcot, Oxfordshire OX11 7HH. Telephone: (44) 01235 400555. Email: primary@hachette.co.uk. Lines are open from 9 a.m. to 5 p.m., Monday to Friday.

Parents, Tutors please call: **(44) 02031 226405** (Monday to Friday, 9:30 a.m. to 4.30 p.m.). Email: parentenquiries@galorepark.co.uk

Visit our website at www.galorepark.co.uk for details of other revision guides for Common Entrance, examination papers and Galore Park publications.

ISBN: **978 1 4718 4935 0**

© Hodder & Stoughton 2016

First published in 2016 by Hodder & Stoughton Limited

An Hachette UK Company

Carmelite House

50 Victoria Embankment

London EC4Y 0DZ

Impression number   15 14 13 12 11

Year   2025 2024 2023

Illustrations by Peter Francis.

The following illustrations are by Integra Software Services Ltd: p.9 bottom, p.33 bottom, p.35 bottom, p.40, p.41 top, p.54, p.55, p.A3 bottom, p.A5 (grids), p.A6 top left, p.A13 right, p.A14

Typeset in India

Printed in the UK

A catalogue record for this title is available from the British Library.

Contents and progress record

Use these pages to record your progress. Colour in the boxes when you feel confident with each skill and note your scores for the 'Test yourself' and workout questions.

C Codes, sequences and matrices

	Completed	Test yourself score

Maths workout 2

	Completed	Score

How to use this workbook

Introduction

This workbook has been written to help you develop your skills in Non-Verbal Reasoning. The questions will help you:

- learn how to answer different types of questions
- build your confidence in answering these types of questions
- develop new techniques to solve the problems easily
- practise maths skills that can help improve your abilities in Non-Verbal Reasoning
- build your speed in answering Non-Verbal Reasoning questions towards the time allowed for the 11+ tests.

Pre-Test and the 11+ entrance exams

The Galore Park 11+ series is designed for Pre-Tests and 11+ entrance exams for admission into independent schools. These exams are often the same as those set by local grammar schools too. 11+ Non-Verbal Reasoning tests now appear in different formats and lengths and it is likely that if you are applying for more than one school you will encounter more than one of type of test. These include:

- Pre-Tests delivered on-screen
- 11+ entrance exams in different formats from GL and CEM
- 11+ entrance exams created specifically for particular independent schools.

Tests are designed to vary from year to year. This means it is very difficult to predict the questions and structure that will come up, making the tests harder to revise for.

To give you the best chance of success in these assessments, Galore Park has worked with 11+ tutors, independent school teachers, test writers and specialist authors to create this series of workbooks. These workbooks cover the main question types that typically occur in this wide range of tests.

For parents

This workbook has been written to help both you and your child prepare for both pre-test and 11+ entrance exams.

The content doesn't assume that you will have any prior knowledge of Non-Verbal Reasoning tests. It is designed to help you support your child with simple exercises that build knowledge and confidence.

The exercises on the **learning spreads** can be worked through either with your support or independently. They have been constructed to help familiarise your child with how a question type works in order to build confidence in tackling real questions.

The **maths workout** sections are provided to help consolidate learning in related areas of maths.

Working through the workbook

- The **contents and progress record** helps you keep track of your progress. Complete it when you have finished one of the **learning spreads** or **maths workout** sections.
  - Colour in the 'Completed' box when you are confident you have mastered the skill.
  - Add your 'Test yourself' scores to track how you are getting on and to work out which areas you may need more practice in.
- **Chapters** link together types of questions that test groups of skills.
- **Learning spreads**, like the one shown here, each cover one question style.

Have a go

Try these activities to build your skills towards answering the exam-style questions.

Test yourself

Complete a set of exam style questions that includes some challenging problems.

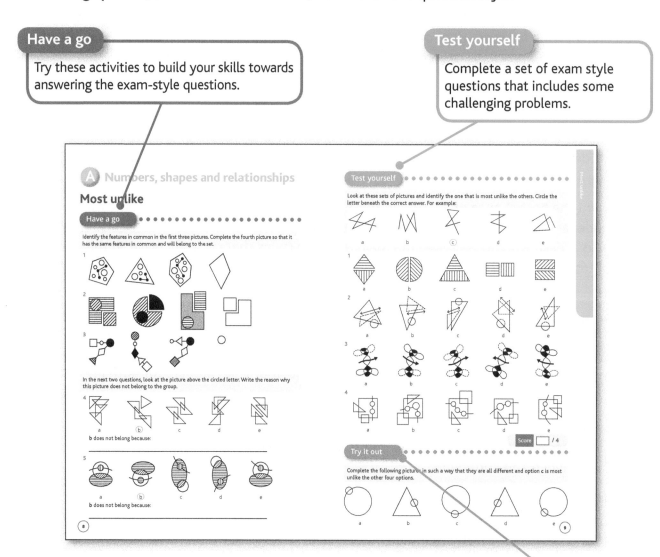

Try it out

Use your new skills to create your own questions or complete a fun activity.

- **Maths workouts** help you to practise familiar skills that link to the Non-Verbal Reasoning questions in this workbook.
- **Answers** to the **Have a go**, **Test yourself** and **Try it out** questions can be found in the middle of the workbook. Try not to look at the answers until you have attempted the questions yourself. Each answer has a full explanation so you can understand why you might have answered incorrectly.

Test day tips

Take time to prepare yourself the day before you go for the test: remember to take sharpened pencils, an eraser and a watch to time yourself (if you are allowed – there is usually a clock present in the exam room in most schools). Take a bottle of water in with you, if this is allowed, as this will help to keep your brain hydrated and improve your concentration levels.

... and don't forget to have breakfast before you go!

Continue your learning journey

When you've completed this workbook, you can carry on your learning right up until exam day with the following resources.

 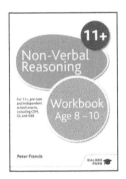 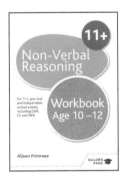  

The Revision Guide (referenced in the answers to this book) reviews basic skills in all areas of nonverbal reasoning, and guidance is provided on how to improve in this subject.

The workbooks will develop your skills with many practice questions. To prepare you for the exam, these books include even more question variations that you might encounter. The more question types you practise, the better equipped for the exams you'll be. All the answers are explained fully.

Workbook Age 8–10: Increase your familiarity with variations in the question types.

Workbook Age 10–12: Develop fast response times through consistent practice.

The *Practice Papers* (books *1* and *2*) contain four training tests and nine model exam papers, replicating various pre-test and 11+ exams. They also include realistic test timings and fully explained answers to help your final test preparation. These papers are designed to improve your accuracy, speed and ability to deal with a wide range of questions under pressure.

Most unlike

Have a go •

Identify the features in common in the first three pictures. Complete the fourth picture so that it has the same features in common and will belong to the set.

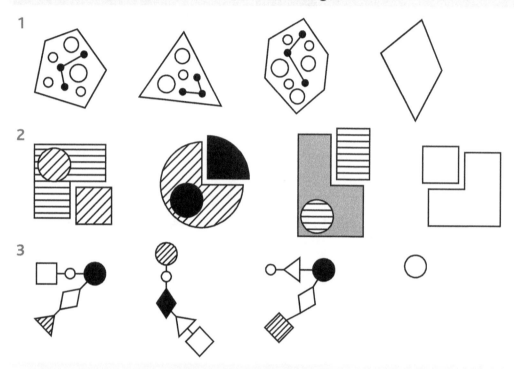

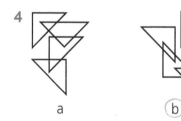

In the next two questions, look at the picture above the circled letter. Write the reason why this picture does not belong to the group.

4

a   ⓑ   c   d   e

b does not belong because:

_____

5

a   ⓑ   c   d   e

b does not belong because:

_____

Test yourself

Look at these sets of pictures and identify the one that is most unlike the others. Circle the letter beneath the correct answer. For example:

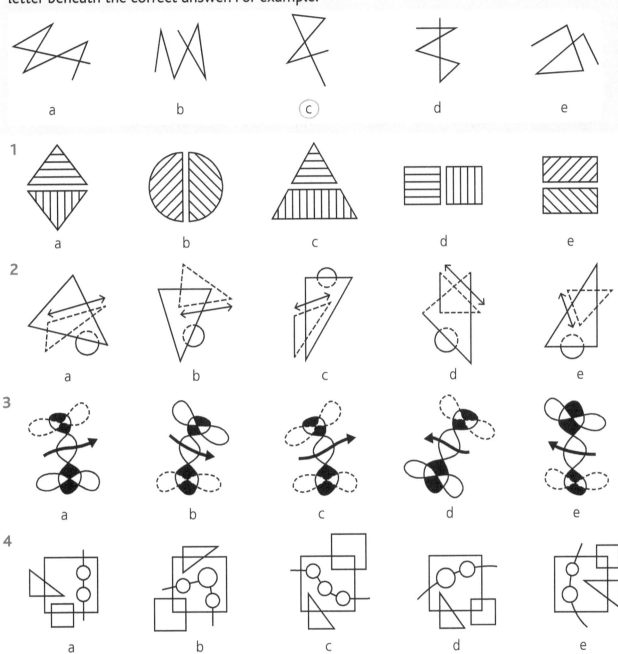

a   b   ⓒ   d   e

1

a   b   c   d   e

2

a   b   c   d   e

3

a   b   c   d   e

4

a   b   c   d   e

Score [   ] / 4

Try it out

Complete the following pictures in such a way that they are all different and option **c** is most unlike the other four options.

   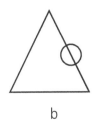   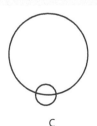      

a   b   c   d   e

Matching features 1

For each set of pictures below, list four things in common within each set.

1

The features in common are:

(a) _____

(b) _____

(c) _____

(d) _____

2

The features in common are:

(a) _____

(b) _____

(c) _____

(d) _____

3

The features in common are:

(a) _____

(b) _____

(c) _____

(d) _____

In the pictures below, options **a**, **b** and **c** form a set. Options **d** and **e** do not belong to the set. Write the feature that means **d** does not belong to the set and the feature that means **e** does not belong to the set.

4

    a   b   c   d   e

_____

_____

5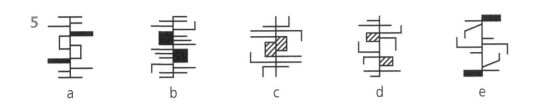

    a   b   c   d   e

Test yourself

Look at the first three pictures and decide what they have in common. Then select the option from the five on the right that belongs to the same set. Circle the letter beneath the correct answer. For example:

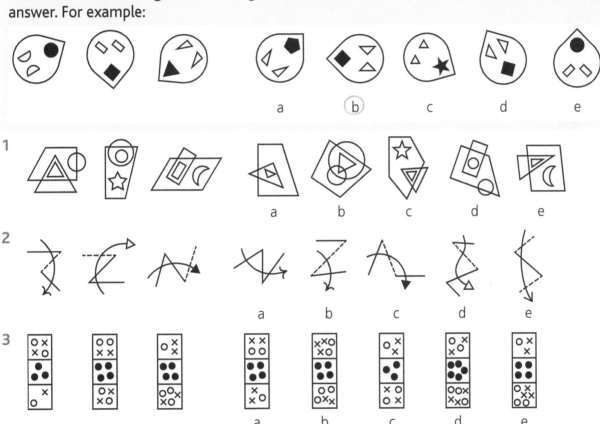

  a   (b)   c   d   e

1

  a   b   c   d   e

2

  a   b   c   d   e

3

  a   b   c   d   e

4

  a   b   c   d   e

Score ☐ / 4

Try it out

Draw your own shapes in boxes **a**, **b** and **c** so that two of your figures belong to the set on the left and one does not. Ask a friend or parent to identify the odd one out.

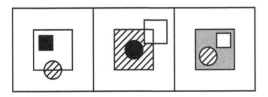 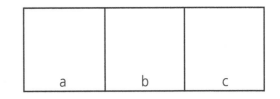

                                                   a   b   c

Applying changes 1

For each question, apply the changes shown in the first chain of shapes to complete chains **(a)** and **(b)** in the same way.

1 → → →

(a) → → →

         i   ii   iii

(b) → → →

         i   ii   iii

2 → → →

(a) → → →

         i   ii   iii

(b) → → →

         i   ii   iii

3 →  → →

(a) → → →

         i   ii   iii

(b) → → →

         i   ii   iii

Test yourself

Look at the two pictures on the left connected by an arrow. Decide how the first picture has been changed to create the second. Now apply the same rule to the third picture and circle the letter beneath the correct answer. For example:

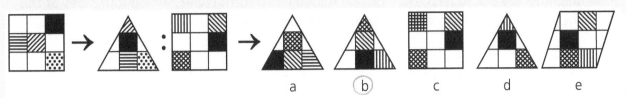

1

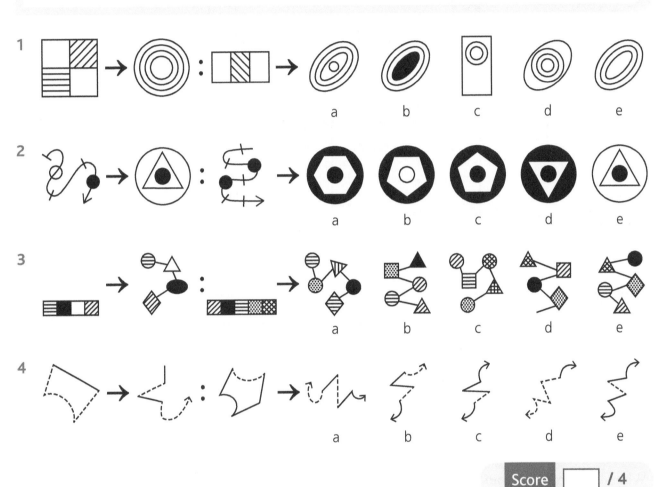

Score □ / 4

Try it out

Complete these questions by creating five possible answer options for each one. Then ask a friend or parent to try them. You can add more detail to the shapes provided if you wish.

Matching 2D and 3D shapes 1

Have a go •

These diagrams are called nets. They can be folded up to form cubes. Which four sides of the cube, when folded, will share an edge with the shaded face?

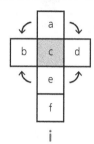

1

	a	
b	c	d
	e	
	f	

i

	a	
b	c	d
	e	
	f	

ii

	a	
b	c	d
	e	
	f	

iii

	a	
b	c	d
	e	
	f	

iv

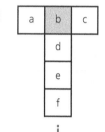

2   i   ii   iii   iv

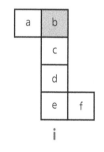

3   i   ii   iii   iv

When these nets are folded to give cubes, the faces with black spots are opposite each other in three of the four cubes. Circle the one where the black spots will **not** be opposite each other.

4

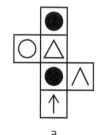
a

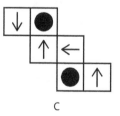
b

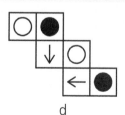
c

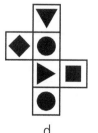
d

5

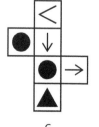

a   b   c   d

14

Test yourself

Find the face that would appear **opposite** the face given on the left when the net is folded into a cube. Circle the letter beneath the correct answer. For example:

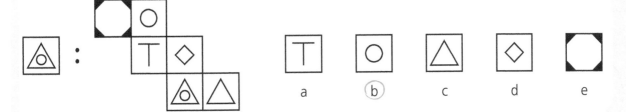

a (b) c d e

1

 :

a b c d e

2

 :

a b c d e

3

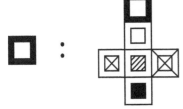 :

a b c d e

4

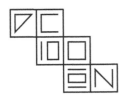 :

a b c d e

Score [ ] / 4

Try it out

Draw a cube net of your own on squared paper. How many different ways can you arrange three circles and three crosses on the net? Identify the nets that will have a circle on opposite faces.

Matching features 2

What **three** elements do these sets of shapes have in common?

1
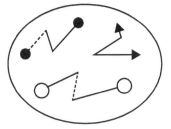

(a) _____

(b) _____

(c) _____

2
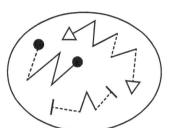

(a) _____

(b) _____

(c) _____

3
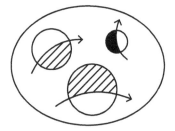 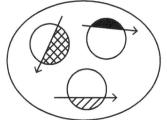 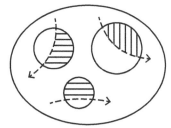

(a) _____

(b) _____

(c) _____

The picture on the left belongs to two of the three groups on the right. Circle the letter of the group to which it does **not** belong.

4
 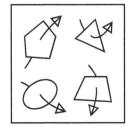  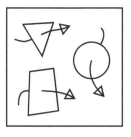

a   b   c

A Numbers, shapes and relationships

5

 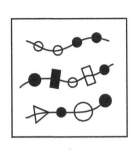 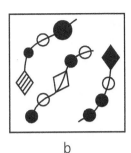 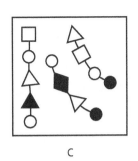

a   b   c

Test yourself

Look at the shapes in the box on the left and decide what they have in common. Then select the option that is part of the same set. Circle the letter beneath the correct answer. For example:

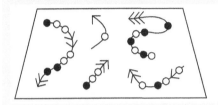

a   (b)   c   d   e

1

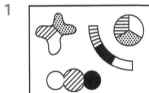

a   b   c   d   e

2

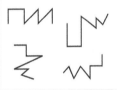

a   b   c   d   e

3

a   b   c   d   e

4

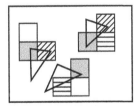

a   b   c   d   e

Score ☐ / 4

Try it out

On a separate piece of paper, draw five more shapes. They should all look very similar to the shapes in the set but only one of them should actually belong to the set. Ask a friend or parent to identify the one that belongs.

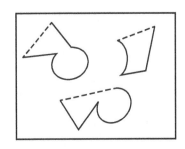

Applying changes 2

1   Identify what changes have been made to the central picture that has resulted in each of the pictures around it. Write what has changed.

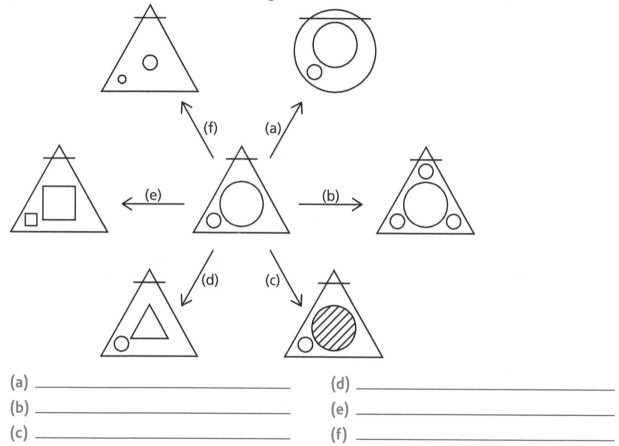

(a) _____   (d) _____

(b) _____   (e) _____

(c) _____   (f) _____

2   Using the instructions written next to the arrows, change the central picture to give six new pictures around it.

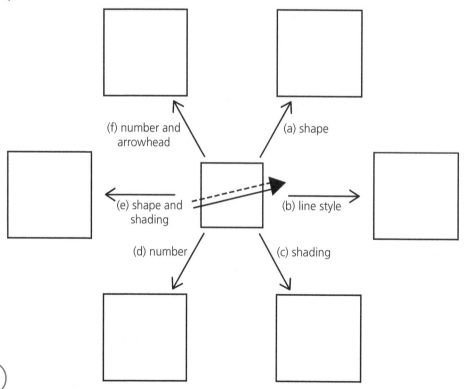

(f) number and arrowhead

(a) shape

(e) shape and shading

(b) line style

(d) number

(c) shading

Test yourself

Look at the two pictures on the left connected by an arrow. Decide how the first picture has been changed to create the second. Now apply the same rule to the third picture and circle the letter beneath the correct answer. For example:

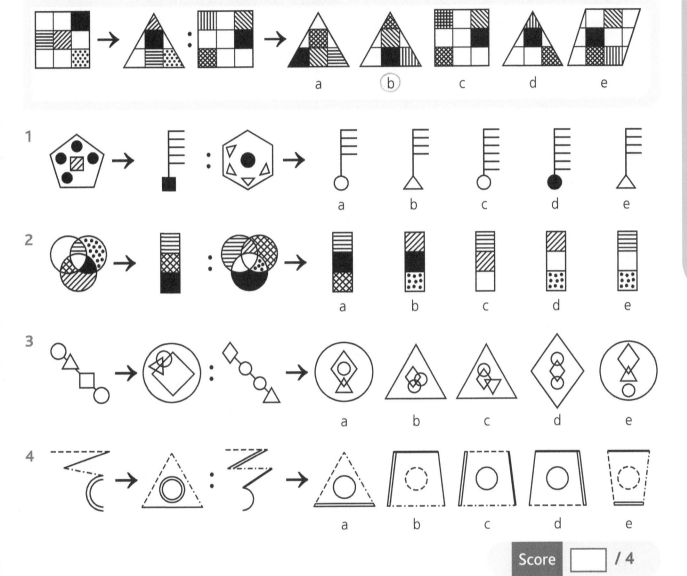

Try it out

Draw a second pair of patterns to go with the first pair, where the first part of the second pair is changed in the same way as the first part of the first pair.

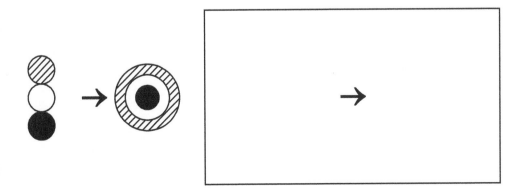

Matching 2D and 3D shapes 2

Draw the 2D plan of these sets of cubes on the squared grid next to them.

1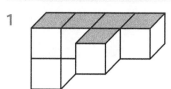

2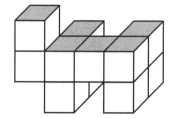

3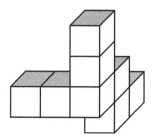

4 Identify the sets of cubes that have the same 2D plan. Write the letters of the pairs and then draw their 2D plan in the grid provided.

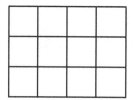 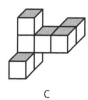 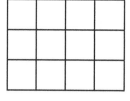 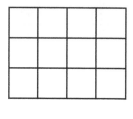
a   b   c   d   e   f

1st pair: _____

2nd pair: _____

3rd pair: _____

Test yourself

Which of the answer options is a 2D plan of the 3D picture on the left, when viewed from above? Circle the letter beneath the correct 2D plan. For example:

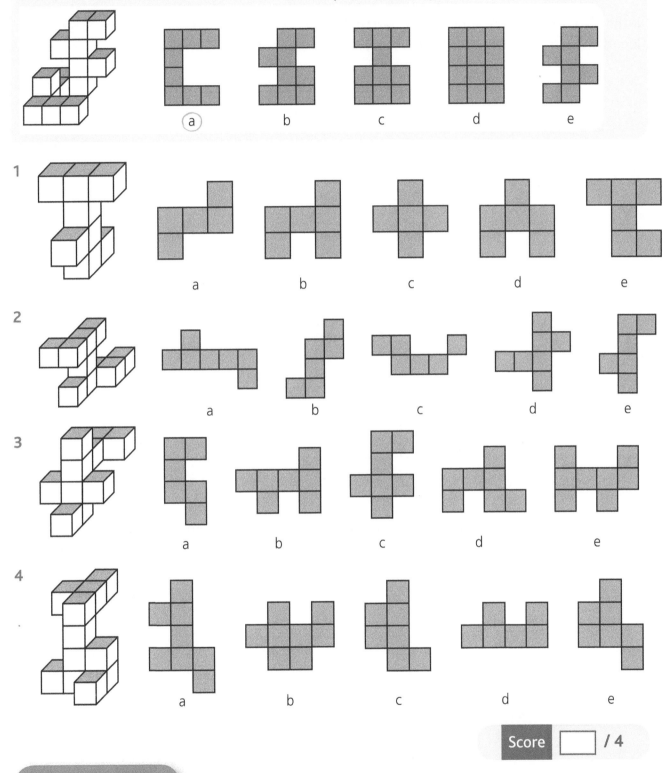

1

2

3

4

Score ☐ / 4

Try it out

How many different 2D plans can be made using four cubes, if each cube has at least one full face touching another cube? Draw them out on a piece of squared paper.

Following the folds 1

If the plan on the left is folded along certain lines it will appear like the diagram on the right. Identify where the folds have been made and draw them in with dashed lines.

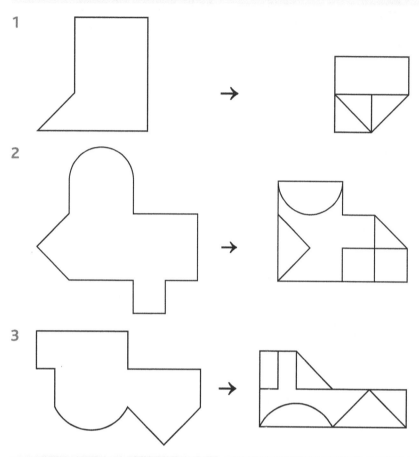

In the next two questions, the diagrams on the left can be made from different plans depending on where the folds are made. Draw two possible plans for each diagram showing the folds with dashed lines.

4 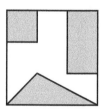 (a)   (b)

5  (a)   (b)

Test yourself

Identify the diagram that shows how the plan on the left will appear when it is folded in along the dashed lines. Circle the letter beneath the correct answer. For example:

a   (b)   c   d   e

1   a   b   c   d   e

2   a   b   c   d   e

3   a   b   c   d   e

4   a   b   c   d   e

Score [   ] / 4

Try it out

On a separate piece of paper, draw your own plan using squares and triangles and dashed fold lines. Ask a friend or parent to draw what it will look like when folded. Cut it out and fold it up to check the answer!

Matching a single image 1

In the next two questions, the picture on the left has been rotated to give the pictures on the right. The rotations are in 45° steps.

i By how many degrees has the picture been rotated clockwise to give each of the pictures on the right?

ii By how many degrees has the picture been rotated anticlockwise to give each of the pictures on the right?

Write the number of degrees in the spaces provided.

1

a

b

  i clockwise   _____   _____

  ii anticlockwise   _____   _____

2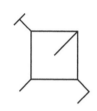

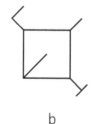

a

b

  i clockwise   _____   _____

  ii anticlockwise   _____   _____

In the next two questions, work out how the picture on the left would appear if it was rotated by the number of degrees indicated and in the direction shown. Draw how it would look in the box provided.

3

225°

90°

4

180°

135°

Test yourself

The picture on the left is rotated as shown by the arrow. Which answer option shows the picture after the rotation? Circle the letter beneath the correct answer. For example:

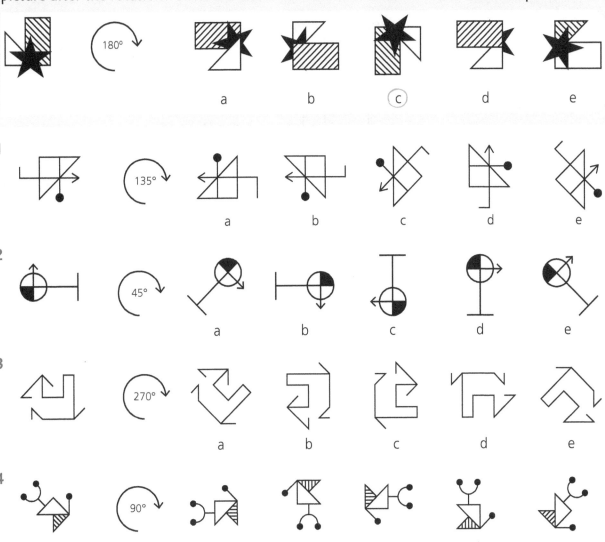

a   b   c   d   e

Score ☐ / 4

Try it out

Draw five answer options for the picture below. Only one of the options should show a 135° clockwise rotation of the picture given. Ask a friend or parent to identify the correct answer.

Translating and combining images 1

1 Which one of the pieces is **not** needed to complete the puzzle?

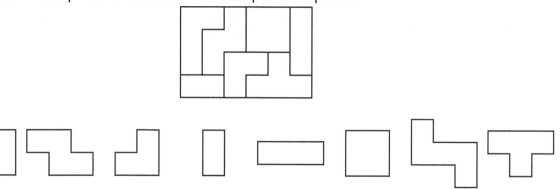

a   b   c   d   e   f   g   h

Fit the shapes into the grid in the next two questions. The shapes may be rotated but not flipped over (that is, not reflected).

2

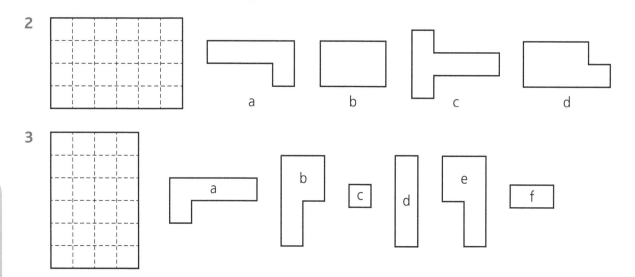

3

Which two shapes on the right are **not** needed to complete the shape on the left?

4

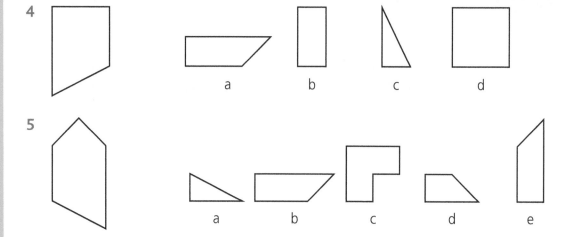

a   b   c   d

5

a   b   c   d   e

Test yourself

The shape on the left can be made using three of the five smaller shapes on the right.
Identify the **three** shapes needed and circle the letters beneath them. For example:

1

a   b   c   d   e

2

a   b   c   d   e

3

a   b   c   d   e

4

a   b   c   d   e

Score ☐ / 4

Try it out

Draw five shapes, four of which can be put together to make the grid below. Ask a friend or parent to identify the one piece that is not needed.

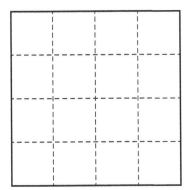

Following the folds 2

Draw the pattern of holes that will be seen when these folded sheets of paper are unfolded.
The fold lines are shown as dashed lines to help you.

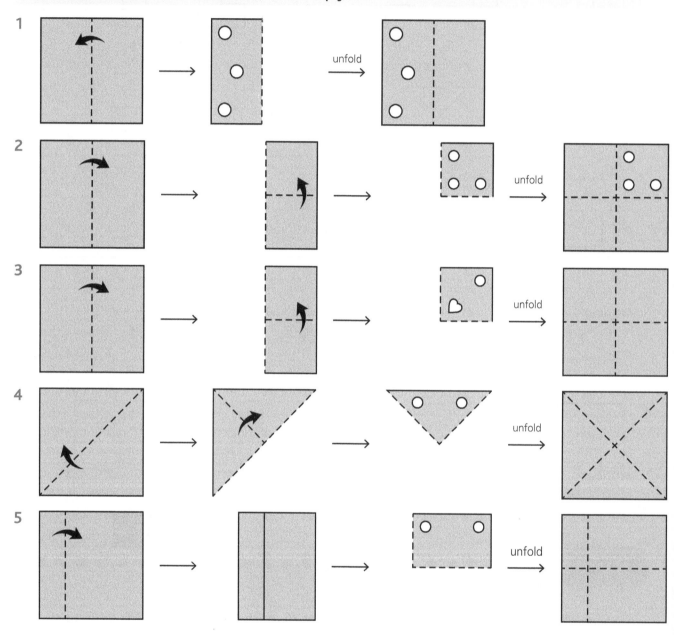

Answers

Please note that all questions are worth one mark unless stated otherwise in brackets.

A Numbers, shapes and relationships

Most unlike (page 8)

Have a go

1   The number of sides of the outer shape matches the number of white circles inside it. Each picture has three black circles connected by a line. The fourth picture should be drawn with four white circles inside the quadrilateral and three black circles connected by a line. Distractors: size of circles does not matter.

2   Each large shape has a circle in the corner opposite the cut-off corner and this circle and cut-off corner are in the same shading. The fourth picture should show a circle in the bottom-right corner of the L-shape, shaded in same way as the separate square top left, with the large L-shape shaded in different style.

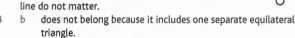

3   Each picture has a small circle, a larger circle, a square, a triangle and a diamond, all connected by a line in any order. The central shape of the five is shaded black and one end shape is shaded with diagonal lines.
Distractors: The order of shapes and the angle of the line do not matter.

4   b   does not belong because it includes one separate equilateral triangle.

5   b   does not belong because the shading of the small circle is horizontal rather than vertical.

Test yourself

1   c   **shape** – the two parts of the shape are not equal size
Distractors: **shading** – inside the shapes; **shape** – outline shape

2   d   **line style** – only large triangle with one dashed side
Distractors: **line style** – of small triangle; **size** – (a) of triangles, (b) of circles; **position** – (a) of circle on large triangle, (b) of arrow

3   c   **line style** – both figure-of-eight shapes at the ends of the picture have dashed lines
Distractors: **size** – of figure-of-eight shapes; **shape** – arrow style; **shading** – of elements of the figure-of-eight shapes

4   a   **angle** – two corners of the small square are inside the larger square, whereas there is only one corner inside in the others
Distractors: **number** – of circles; **orientation** – (a) of line with white circles, (b) relative position of the quadrilaterals and triangles; **line style** – of 'string'

Try it out

Additional shapes and/or shading with one feature of option c making it different from the others.

Matching features 1 (page 10)

Have a go

1   Accept any four sensible answers, which could include the following:
All use the same line style for the three shapes within each figure
All have a circle as the innermost shape
All have a circle as the outermost shape
All have diagonal line shading on inner circle
In all of them the outermost circles are the same size

2   Accept any four sensible answers, which could include the following:
All have three lines making a zig-zag
All have solid lines for the zig-zag
All have a solid line in a curve
All have a circle crossing the curved line
All have half the circle shaded black
All have a short straight line across one end of the curved line

3   Accept any four sensible answers, which could include the following:
All within a curved irregular shape
Outline shape is always a solid line style
All have two white shapes within the outline
One of the inner shapes is half the other
Inner shapes have a solid line style
All have the same number of black circles as there are short lines on the curved line

4   d   the two angles along the zig-zag line are outside the triangle
e   the triangle has two dashed lines

Test yourself

1   e   **shape** – (a) outline shape is a quadrilateral, (b) shape of intersecting shape across side of quadrilateral is the same as the shape inside it; **number** – there are three elements in addition to the quadrilateral

2   c   **number** – zig-zag made up of three parts; **line style** – one part of zig-zag is a dashed line; **shape** – a C-shaped curve crosses the zig-zag in three places
Distractors: **size** – length of zig-zag sections; **shape** – arrowhead style; **proportion** – of lines and angles within the zig-zags

3   d   **number** – (a) total of white circles is same as number of black spots, (b) total number of crosses is same as number of black spots
Distractors: **proportion** – (a) of crosses in first and third squares, (b) of white circles in first and third squares

4   d   **size** – one half of circle shaded
Distractors: **number** – of sections within the circle; **size** – of the sections within each circle; **shading** – style of shading of sections

Try it out

For example, **b** has a rectangle and a square rather than two squares.

Applying changes 1 (page 12)

Have a go

1   For example:
i   step one increases the number of sides of the polygon
ii   step two adds diagonal shading
iii step three adds double line round the shape

2   For example:
i   step one adds a smaller version of the shape in the centre and shade central small shape only
ii   step two adds another small white shape overlapping the bottom of the main shape
iii step three adds another small shaded shape touching the outside of the main shape in the top-right quarter

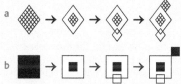

3   For example:
i   step one draw a zig-zag line across the shape with the same number of sections as there are black circles
ii   step two add black circles to ends and angles of zig-zag line
iii step three add curved line plus white circle

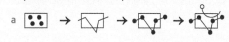

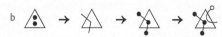

Test yourself

1   e   **number** – three rectangles will give three concentric shapes; **shape** – rectangles give ovals; **shading** – none in the second shape
Distractors: **shading** – of elements in the first shape

2   c   **number** – the number of lines across the arrow gives the number of sides of the polygon; **shape** – five lines gives a pentagon; **size** – one circle increases in size; **shading** – both circles are black; **position** – the smallest and largest shapes are circles and the shapes sit inside each other

3   e   **number** – the same number of small shapes as squares; **shading** – one shape with each style of shading
Distractors: **shape** of elements; **order** – (a) of shapes, (b) of shaded elements

4   d   **number** – same number of straight and curved lines in second picture as in first picture; **line style** – (a) straight lines change from solid to dashed or vice versa, (b) curved lines remain same style; **shape** – arrowhead added at open end of curved lines
Distractors: **size** – length of lines

Try it out

1   Answer should be a quadrilateral with three small white circles and one cross inside.

2   Answer should show the line styles clockwise around a quadrilateral in the same order as the line styles given.

Matching 2D and 3D shapes 1 (page 14)

Have a go

In the first three questions all the faces will share an edge with the shaded face **except** the face that will be on the opposite side.

1   i abde   ii bcdf   iii acef   iv bcdf
2   i acdf   ii abce   iii bdef   iv acdf
3   i acef   ii abdf   iii bcde   iv abdf

In the next two questions the faces that share an edge cannot be opposite. The two edges that will be next to each other when the net is folded are shown with a bold line.

4   d 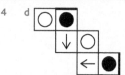   5   c

Test yourself

1   c   Once the net is folded, the opposite face is the only one that does not touch the original face. The white triangle, plus sign and double circle are already touching the face and the black spot will touch the given face when folded. The face with the black triangle is unable to touch the given face and so is the opposite face.

2   a   Once the net is folded, the opposite face is the only one that does not touch the original face. The T-shape is already touching the face, the O-shape, the inverted V-shape and the C-shape will fold around and so touch the given face. The face with the U-shape is unable to touch the given face and so is the opposite face.

3   d   Once the net is folded, the opposite face is the only one that does not touch the original face. The white square in a white box is already touching the face, the two squares with crosses and the smaller black square will also touch when folded up. The diagonally shaded square in the white box is unable to touch the given face and so is the opposite face.

4   e   Once the net is folded, the opposite face is the only one that does not touch the original face. The white square is already touching, and the rectangle with one line, the Z-shape and the triangle will fold around and touch the given face. The face with the U-shape is unable to touch the given face and so is the opposite face.

Try it out

There are 20 possible permutations. Of these, 12 will have a pair of circles on opposite faces. The following is one example only – there are many other nets that could be drawn.

Matching features 2 (page 16)

Have a go

1   Accept any three sensible answers, including:
Made up of zig-zag shapes
Solid and dashed lines
Shapes at the end of each zig-zag are the same

2   Accept any three sensible answers, including:
All patterns made up of a triangle and a quadrilateral
Triangles are shaded
Shading style same throughout one set

3   Accept any three sensible answers, including:
All have three circles
All have an arrow crossing each circle
All arrowhead styles are the same
One section of each circle is shaded

4   b   All shapes in this set have the arrow passing through them from the front to the back

5   c   These strings are made up of more than two different shapes and do not have two black circles along them

Test yourself

1   d   **number** – divided into three sections; **shading** – different shading styles used for two sections with one left white
Distractors: **shading** – styles used are unimportant; **shape** – the outline shape is unimportant; **size** – the size of the sections is unimportant

2   e   **number** – (a) three lines form a zig-zag with two right-angles before continuing, (b) four more lines form a zig-zag with irregular angles
Distractors: **size** – line length of each section is unimportant

3   c   **position** – (a) V-shapes have opening next to a solid line and away from a dashed line, (b) C-shapes have opening next to a dashed line and away from a solid line
Distractors: **size** of V- and C-shapes is unimportant; **number** – of V- and C-shapes is unimportant; **position** – of V- and C-shapes along the line is unimportant

4   d   **number** – set of three adjoining squares; **position** – one triangle overlapping the set of squares; **shading** – (a) of squares can be seen through the triangles, (b) triangle is unshaded
Distractors: **number** – (a) of squares overlapped by the triangle, (b) of squares shaded; **shading** – style of shading used for the squares

Try it out

There are many possible answers. For example:

| a | b | c | d | e |

Applying changes 2 (page 18)

Have a go

1. (a) shape – outer triangle to circle
   (b) number – one small circle to three small circles
   (c) shading – large central circle white to shaded
   (d) shape – central circle to a triangle
   (e) shape – both shapes inside triangle from circle to square
   (f) size – central shapes from large to small

2. Many answers possible, for example:

| (a) | (b) | (c) |

| (d) | (e) | (f) |

Test yourself

1. c **number** – six short horizontal lines from the upright, same as number of sides of the polygon; **shape** – at base is same as the shape in the centre of the polygon; **shading** – at base has same shading as the small shapes around the inside edge of the polygon
   Distractors: **number** – of small shapes is not relevant

2. d **shading** – (a) top square shading same as the overlapping section of the top two circles, (b) middle square same shading as the overlapping section of the two circles on the left, (c) bottom square same shading as the overlapping section of the two circles on the right
   Distractors: **shading** – styles of the other sections

3. b **shape** – (a) same shapes are used, (b) the outer shape is one of the shapes at the ends of the string; **relative position** – (a) three of the shapes overlap within the fourth shape; (b) overlapping shapes all within a large shape
   Distractors: **relative position** – of the overlapping shapes

4. c **number** – outer shape has the same number of lines as the lines in the zig-zag; **shape** – a circle inside the polygon; **line style** – (a) same styles of the four lines in zig-zag around the quadrilateral, (b) line style of the curved line is used for the circle
   Distractors: **length of lines** – in the zig-zags

Try it out

There are many possible answers. For example:

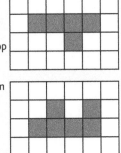

Matching 2D and 3D shapes 2 (page 20)

Have a go

1. The upper layer of four cubes gives the shape of the plan as the single cube on the lower layer is directly beneath the top left-hand cube and so will not show on the plan (if you look at it from above).

2. The middle layer of cubes gives the pattern of the plan as the cubes on the top and bottom layer are all directly above or beneath a cube in the middle layer.

3. The middle layer of cubes gives an L-shape where the vertical part of the L is two cubes and the horizontal section is three cubes long. The cubes in the two upper layers sit directly above cubes in the middle layer and so do not affect the plan. The cubes in the bottom layer extend for one cube beyond the middle layer on the right side, extending the vertical part of the L-shape, so that both parts are three cubes long in the plan.

4. Pair 1: **a** and **f**; Pair 2: **b** and **c**; Pair 3: **d** and **e**

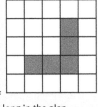

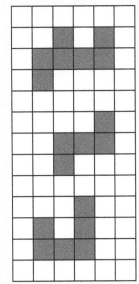

Test yourself

1. c The bottom layer has two cubes in a line. This line is extended forward by the one cube on the second layer to become three in a line. The third layer does not extend beyond this, but the fourth layer extends by one cube each side across the central cube, making the plus shape plan of option **c**.

2. b The bottom layer has three squares in a line with two further squares on the right of one end, overlapping by one square and longer by one square. The middle layer single cube does not affect the plan. The top layer covers the original three cubes in a line and has a single cube projecting to the left side at the front end of the row, giving plan **b**.

3. c The bottom layer of two cubes gives a line that is extended back by another two cubes in the third layer, making a line of four. The second layer gives a cube on each side that is just one cube along from the end. The furthest end of the third layer has a single cube projecting to the right. The single cube in the fourth layer does not extend further, giving the plan in option **c**.

4. e The bottom layer gives three cubes in a row with one projecting forward at the right-hand end. The cubes in the second and third layers sit over these cubes. In the fourth layer two cubes project backwards from the central cube of the bottom row of three and an additional cube projects left from the central one of these three, giving the plan in option **e**.

Try it out

There are eight possibilities as shown here:

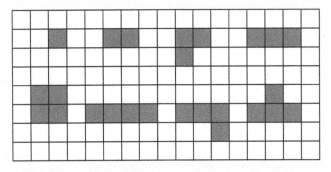

Following the folds 1 (page 22)

Have a go

1 The triangle in the lower left corner of the diagram opens out to the left so a dashed line extends down to the bottom of the vertical left side. The lower right triangle opens out, so there is a dashed line diagonally in the lower right corner. The rectangle across the top opens up so needs a horizontal dashed line.

2 The semi-circle is folded in so there is a dashed line across the top of the plan beneath the semicircle. The straight vertical line on the left extends down with a dashed line as the triangle on the left folds in. A dashed fold line goes across the single square projecting down from the bottom edge. There is a diagonal fold line from lower right to upper left across the square corner on the right.

3 There should be dashed fold lines as shown.

4 For example, where a shaded section (representing a part folded in) shares two straight edges with the outside of the diagram, then there are two possible positions in the plan for that shape. So here the square could open up along the top edge or the left side and the rectangle could open up along the top edge or the right side. So there are four possible combinations.

5 For example, where a shaded section (representing a part folded in) shares two straight edges with the outside of the diagram, then there are two possible positions in the plan for that shape. So here the square could open up to the top or to the right and the trapezium can open up to the right or to the bottom. So there are four possible combinations.

Test yourself

In each of these the dashed fold lines act as lines of reflection for each small shape, so the mirror image of each small shape will be shown inside the larger shape.

1 c   2 b   3 d   4 e

Try it out

There are many possible answers.

Matching a single image 1 (page 24)

Have a go

1 (i) a 45°   2 (i) a 225°
    b 90°   b 180°
  (ii) a 315°   (ii) a 135°
    b 270°   b 180°

3   4

Test yourself

1 c 135° is one and a half right-angles. Rotating the figure clockwise by 45° gives option e, by 180° gives option a, by 90° anticlockwise gives option d and option b is a reflection

2 e 45° is half a right-angle. Rotating the figure by 90° gives option d, by 180° gives option b, by 135° gives option a, by 270° gives option c

3 b 270° is three right-angles or a three-quarter turn. Rotating the figure by 45° gives option a, by 90° gives option c, by 180° gives option d and 225° gives option e

4 e 90° is one right-angle. Rotating the figure 315° gives option a, by 225° gives option b, by 135° gives option c and 45° gives option d

Try it out

The correct answer must include this option:

Translating and combining images 1 (page 26)

Have a go

1 g is not needed – working across the top half of the rectangle are pieces a and b both rotated through 90°, then f and then e which is also rotated; on the lower part is d rotated through 90°, and then c and h both rotated through 180°

This is one example only – there are other possibilities.

2 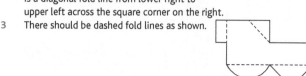   3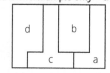

4 a and b not needed   5 c and d not needed

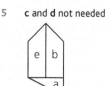

Test yourself

1 a c d   2 a b c

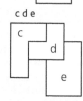

3 a b d   4 c d e

Try it out

There are many possible answers. For example, b is not needed to complete this grid:

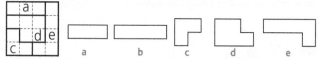

Following the folds 2 (page 28)

Have a go

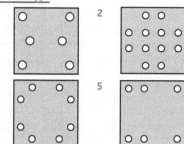

1   d   The fold lines act like lines of reflection. When unfolding the second fold there will be two circles along the top of the rectangle, one at each end as it will be a reflection of the rectangular shape in the top section of the final folded piece. Unfolding the first fold to the left will give three circles in a vertical line.

2   e   The fold lines act like lines of reflection. Note that the heart shape will point towards the top of the square.

3   c   The fold lines act like lines of reflection. Note that the two vertical holes are not in the centre of the vertical line.

4   c   The fold lines act like lines of reflection. So, when unfolding the second fold, a right-angled triangle will be mirrored from the top-right corner of the lower left square into the lower right corner on the top (which still appears as a triangle). When the first diagonal fold is unfolded there will be a pair of right-angled triangles midway along the top edge of the large square, and four circles in the lower left smaller square, as reflected in a diagonal mirror line.

Try it out

The order of folding can vary but needs to end up with a small, vertical rectangle that is one-eighth of the square and has one circle punched out in the top half. For example:

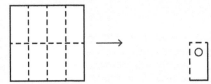

B Position and direction

Matching a single image 2 (page 30)

Have a go

1   c and d   The easiest way to check is to hold a mirror vertically halfway along the picture, dividing it into two parts. The reflection will match the other part of the picture if it is reflected accurately in a vertical mirror line.

2

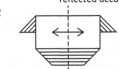 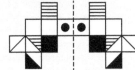

3   (a)   A H I M O T   (b)   B C D E H I O

Test yourself

1   c   a is an incorrect picture, b is nearly an exact copy of the question, d has no shading and the dashed line is in wrong position, e has inaccurate proportions

2   d   a and b are not copies of the picture, c has line angle wrong in the top-right corner and an extra short diagonal line, e is nearly a copy of the question

3   e   a, c and d all have some elements of the first picture incorrectly reflected and b is a horizontal reflection correctly reflected, but not accurately

4   e   a and c have the arrow in the wrong position, b is a horizontal reflection, the diagonal line across the rectangle is incorrect in d

Try it out

There are many different possible answers. Check by holding a mirror along the dashed line.

Translating and combining images 2 (page 32)

Have a go

1   (i)   1
     (ii)   1
     (iii)   2

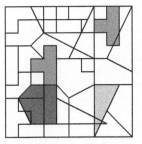

2   c   The rectangles in a are not complete, in b they are too thin, there are no similar rectangles in d

3   b   Looking for a right-angled triangle with the two short sides of equal length within the pattern eliminates a, c and d

Test yourself

1   c   The two triangles are vertically opposite each other and both have right-angles at the point of contact. Look at the bottom left of option c.

2   d   The shape is made up of two pieces – the shape with the semi-circle on the right and the triangle.

3   d   The most identifiable element of this long thin triangle is that the sharp 'point' vertex is not vertically above the base line.

4   b   The shape to find is a triangle with the tip cut off, or it can be envisaged as a triangle next to a small quadrilateral. The triangles in the other options lean the wrong way or do not match the shape exactly.

Try it out

There are many possible answers. For example:

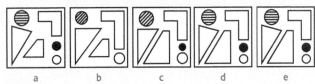

Matching a single image 3 (page 34)

Have a go

1   i f   ii c   iii d   iv e
2   i f   ii a   iii b
3   b
4   c
5   b   Note direction of stripes

Test yourself

1   a and d   **line length** – (a) the two end lines of the Z-shape are short, (b) the central line is long; **shading** – (a) one circle at one end is black, the other white, (b) central circle is white; **orientation** – a and d are rotations of the same picture and are therefore identical, but e is a reflection

2   b and e   **number** – six line segments; **rotation** – e is b rotated through 180°

3   b and c   **number** – two small circles at each end of arc; **shading** – (a) both black at one end and white at the other end, (b) lines across the middle-sized circle; **position** – smaller black circle inside the line-shaded circle is touching the outer line at the top left

4   a and e   **number** – (a) of horizontal lines, (b) of diagonal lines from central vertical line at base; **position** – of horizontal lines

Try it out

There are many possible answers. For example, d and e are identical here.

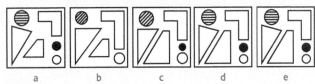

Translating and combining images 3 (page 36)

Have a go

1   i g   ii e   iii a   iv c   v h
2   i   c and e   ii b and d   iii a and f
3 a   4 d   5 c

Test yourself

1 a One small right-angled triangle removed. In **c** the quadrilateral is too large. The triangle in **d** is not right-angled.

2 c The original shape is made up of a 'house' resting on a parallelogram, and an isosceles triangle is removed. Option **a** is a mirror image of the original. Some has been added to **b**. In **d** and **e** two triangles have been cut away. In **c** the triangle has been removed from one side of the parallelogram.

3 c In option **a** an additional triangle has been removed from the top. **b** has the triangle removed from the left-hand side but the pattern of triangles on the base is not the same as the original. Options **d** and **e** have small right-angled triangles cut away or added.

4 b Focus on the small rectangle, which, in **b**, has been cut out of the shape in the same orientation as the original. In **a** and **d** a small rectangle has been cut away, but the original smaller rectangular cut has been lost. Option **c** is the same as the original shape with no rectangle cut away and **e** is a different shape.

Try it out

There are many possible answers. For example, here **i** has a square cut away and **ii** has a triangle cut away.

i   ii

Translating and combining images 4 (page 38)

Have a go

1

Pattern	Block A	Block B	Block C	Block D
i	1	2	0	0
ii	1	1	1	0
iii	0	2	0	1
iv	1	2	0	1

2 b   3 a   4 c

Test yourself

1 c The L-shaped block is rotated 90° clockwise in a horizontal plane, with the front end resting on a cube that is placed next to the other cube. The cuboid is balanced on the top, extending over the cube beneath but not sitting on it.

2 e One of the L-shaped blocks is rotated into a horizontal position and fitted into the other. The end of the upper L-shape is resting on the cuboid.

3 e One of the long cuboids is sitting on top of the inverted L-shaped block, and the other is beneath it at right-angles and resting on the single cube

4 a The T-shaped block is turned upside down and the L-shaped block fits round one side. The long cuboid sits along the base next to the horizontal bar of the T-shape.

Try it out

There are many possible answers. Here are three examples.

Maths workout 1 (page 40)

Working with 2D shapes

1 a b e f   2 b c d e
3 i (2,0)   ii (3,1)   (2 marks)
4 i (3,3)   ii (2,0)   (2 marks)

Working with 3D shapes

1 a 38 cm² b 28 cm² c 34 cm² d 38 cm²   (4 marks)
2 (a) i 1   ii 1   (b) i 7   ii 1
                                       (4 marks)

Codes, sequences and matrices
Connections with codes 1 (page 42)

Have a go

1 a Y – the number of circles is determined by the second letter
  b A – the shape is determined by the first letter (the rotation of the shape is a distractor)

2 a S – the shape of the arrowhead is determined by the second letter
  b M – the number of lines making up the arrow is determined by the first letter (the curved line is a distractor)

3 a E – the number of parallel lines is determined by the first letter
  b M – the shading in the circle is determined by the second letter
    Distractor 1: **number** – of sides on large outer shape
    Distractor 2: **position** – of short lines across edge of shape

4 a Z – the number of circles is determined by the second letter
  b B – the number of triangles is determined by the first letter
    Distractor 1: **shape** – triangles
    Distractor 2: **shading** – of circles
    Distractor 3: **position** – of circles

5 a Y – the shading style is determined by the third letter
  b N – the proportion of the shape shaded is determined by the second letter
  c A – the outer shape is determined by the first letter

Test yourself

1 b **orientation** – the first letter represents the position of the shield: A pointing up, B to the right and C down; **number** – the second letter represents the number of lines across the straight edge of the shield: F two lines, G one line
    Distractors: **number** – of points on the star; **shading** – of star

2 c The first letter refers to the **number** of squares in the pattern, A is for 3 and B for 4 squares; the second letter refers to the **shading** of the large circle, P has two lines across it, Q has one line and R is fully shaded; and the third letter refers to the **number** of black dots in one of the squares, X for 1, Y for 2 and Z for 3.
    Distractors: **angle** – of line across white circle; **shape** – formed by the adjoining squares; **orientation** – of the black spots.

3 e the first letter is for the **number** of curved lines: A is for one; the second letter is for the basic **shape** (L, Z or F) so answer is Y for a Z-shape on its side; the third letter is for the **orientation** of the pair of short lines that cross one of the vertical lines, so G is for them crossing at right-angles
    Distractors: **orientation** – of the letter; **line style** – of the pair of short lines

4 d the first letter is for the **number** of small white circles inside the large white circle, so M for three; the second letter is for the **shading** style of the section where the small circle overlaps the large circle: so X for remaining white, so answer is MX
    Distractors: **size** – of smaller circle; **number** – of straight lines across the circle; **position** – of smaller circles across the large circle

Try it out

There are many possible answers. For example:

(a) AMT   (b) ALS   (c) CMR   (d) BNR

where the first letter is for the number of short straight lines across the triangle, the second letter is for the shading of the circle inside the triangle and the third letter is for the shading of the shape outside the triangle.

Sequences 1 (page 44)

Have a go

1 A rectangle made up of two adjacent squares, with one spot in the first square and two in a diagonal line across the second square from top left to bottom right

2 A circle with a solid arrow pointing to the left from a central spot in the circle, with the arrow ending at the circumference

3 A cross should be added to **c**, a small white triangle to **d**, a small white circle to **e**

4 The two circles should be black, the corner at the base of the triangle should be shaded with horizontal lines and the central square will be all black

Non-Verbal Reasoning Workbook Age 9–11 published by Galore Park

5   A small black circle sitting inside the crescent shape, touching but not crossing it; and four short horizontal lines, two coming out to the left from the top and two from the bottom of the crescent should be added

Test yourself

1   c   **position** – the larger circle moves down the string; **number** – (a) of small black spots reduces by one each time, (b) of lines in the solid zig-zag reduces by one each time, (c) of lines in the dashed zig-zag increases by one each time
Distractors: **position** – (a) of the zig-zags within the square, (b) of the point where the dashed zig-zag crosses the thread line

2   d   **number** – (a) of circles decreases by two each time, (b) of shaded segments decreases by one each time
Distractors: **size** of circles; **position** of shaded sections

3   d   **proportion** – shaded section in the circle on the left increases by a quarter each time; **shading** – style of shading in the left-hand circle alternates between diagonal stripes and black; **angle** – (a) the dashed radius line in right-hand circle moves 45° clockwise each time, (b) the solid radius line in right-hand circle moves 90° anticlockwise each time
Distractors: **shading** – style of background

4   d   **position** – (a) the shaded square moves one place round the edge of the larger square anticlockwise each time, (b) the white circle is always directly opposite the shaded square, (c) the black circle is one square ahead of the white circle; **shading** – (a) angle of line shading in the square rotates 45° clockwise each time, (b) the central square alternates white and black

Try it out

Many different solutions are possible. For example:

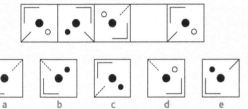

Matrices 1 (page 46)

Have a go

1   The first grid has the same shape in each column and the shading along each row alternates so:
i   **h**   white diamond   ii   **a**   black circle
iii   **e**   white triangle
In the second grid the pattern moves down the grid in a zig-zag way along the top row from left to right and then back across the middle row and along the bottom row. So the shapes go circle, triangle, diamond, circle, triangle, diamond, and so on while the shading goes stripes, black, black, white, stripes, black, black, white, and so on.
iv   **g**   black diamond   v   **f**   striped triangle

2   There is a diagonal line from bottom-left corner to top right, the triangle on the upper left side is divided into two with a line from the top-left corner of the square to the centre of the square. The small triangle on the left has vertical line shading and the small triangle at the top right has horizontal line shading

3   There should be three circles in the square, each one shaded with grid line shading. The number of circles is indicated by the number of crosses in the triangle to their right and their shading is determined by the shading of the triangle beneath them.

4   (a)   shading   (b)   shape   (c)   line
(d)   background shading

5   This grid has four lines of symmetry – one vertically across the centre, one horizontally across the centre, and then two diagonal lines of symmetry also passing through the centre. The box should have a wide V-shape drawn in the lower half with the inner angle of the V shaded black, and two thin parallel lines going across the top of the square. It is a reflection of the top square in the grid.

Test yourself

1   b   the **shapes** alternate down the columns, so it will be a triangle pointing left; **shading** style alternates along the rows so it will have vertical stripes

2   d   **shape** – (a) the symbol in top-left corner of each square is the same along each diagonal line, so there will be a small right-angled triangle across the top-left corner, (b) shape in bottom-left corner is same along each row, so there will be a spade shape; **number** – of black spots on the right same in each row, so there will be three; **shading** – one of each style of shading on each row, so spade will be black

3   e   **number/position** – two shapes, one in top-right corner and one in lower left corner (which is in the middle of the grid); **shape** – (a) the lower left shape in the square is taken from the border of the next square going clockwise, (b) the shape in the upper right corner is from the adjacent border and will be an L-shape; **shading** – (a) the lower left shape is black, (b) the upper right shape is white

4   b   **rotation** – anticlockwise by 90°; **position** – of shaded element of the small circle pattern; **orientation** – of each element in relation to the sides of the square

Try it out

There are many possible answers. For example:

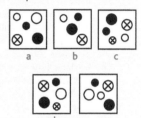

Connections with codes 2 (page 48)

Have a go

1   i   AN   a square with horizontal stripes
ii   BL   a black circle
iii   CM   a white triangle
iv   BM   a white circle
v   CN   a triangle with horizontal stripes
vi   AL   a black square

2   i   EZ   ii   DX   iii   FZ
iv   FY   v   DY

3   A   one triangle
B   two triangles
X   no shading
Y   two shapes shaded
Z   one shape shaded

4   D   zig-zag of three lines
E   zig-zag of two lines
F   zig-zag of four lines
S   circle shaded black
T   circle with diagonal cross line shading

5   Completed picture must include an arrow pointing down and a curved line with two black circles on it – the white circles and arrowhead style are distractors and so do not matter.

Test yourself

1   b   first letter is for **shape** of base; second letter is for **number** of flowers
Distractors: **lines** – presence or absence of leaves; **shading** – of shape at base

2   e   first letter is for **position** of the shaded shape in the line, so R for middle; second letter is for the **sequence** of shapes, so X for rectangle, circle and oval from bottom to top
Distractors: **angle** – of line across the square

3   c   first letter is for **angle** of diagonal dashed line; second letter is for **number** of solid straight lines
Distractors: **orientation** – of straight lines; **angle** – of intersections of straight lines; **size** – of curved lines

4   d   first letter is for the **shape** of the three diagonal items, so P for black dots, Q for white dots and R for crosses; second letter is for **position** of white circle on the top row, L for left and M for right
Distractors: **position/shapes/shading** of the other elements

Try it out

There are many possible answers.

Sequences 2 (page 50)

Have a go

1. a   Small circle with vertical stripes as the lines rotate by 45°
       anticlockwise each time, progressing along the sequence
   b   Three short arrows pointing right, in a vertical line as the
       orientation of the arrows remains the same along the top and
       bottom rows but the number increases by one
   c   Six small black circles arranged in any way in the bottom left corner
       because their number increases with no particular arrangement

2. a   A white circle because alternate L-shapes have a circle at the
       top right that alternates between black and white shading
   b   A short horizontal arrow pointing left because there is an arrow
       at the top of the back-to-front L-shapes, rotating 45° clockwise
       each time
   c   A white equilateral triangle because there is an equilateral
       triangle beneath each arrow alternating between black and white

3. i   c   The number of short lines decreases by one each time so
           there should be four lines, with the centre of the crossed
           lines forming a shape with the same number of sides –
           option **c** has a quadrilateral in the middle
   ii h   The shaded third of the circle moves one position anti-
           clockwise each time and the shading style alternates

4. i   b   Equilateral triangles alternate with right-angled triangles,
           and alternate equilateral triangles are inverted. When
           the 'base' line is across the top it has a solid line outside,
           otherwise this line is dashed
   ii h   Right-angled triangle rotates 90° anticlockwise each time,
           with a solid line outside when 'base' line is across the top

5. Various possible options, for example:
   a   circle with an arrow through it pointing left
   b   equilateral triangle with diagonal stripes from lower left to top right
   c   square with three black circles forming a vertical line down the
       middle

Test yourself

1. e   **position** – (a) U-shape along top line will be at the right-hand side
       of the square as it is moving along, (b) double dashed line across
       the bottom of the square; **shading** – a white circle in the middle
2. d   **angle** – (a) solid line arc of outer circle encompasses 180°,
       (b) solid line arc completes outer circle encompasses 180°,
       (c) inner arc encompasses 90°; **rotation** – (a) outer arrow
       moves 45° anticlockwise each time, (b) inner arrow moves 90°
       clockwise each time; **shading** – (a) of quarter section of central
       circle moves anticlockwise round each time, (b) arrowhead of
       inner arrow alternates from black to white
3. c   **position** – right-angled triangle in top-right corner; **shading** – (a)
       black square bottom left, (b) circle in centre of square has X inside
4. e   **rotation** – arrow rotates 90° anticlockwise each time; **line
       style** – tail of arrow changes from solid to dashed to double
       lines each time; **shading** – small circle at end of line that
       crosses the arrow alternates from black to white; **position** –
       the larger circle that is divided into quarters moves from end of
       line with small circle, to middle, to open end and then back to
       the small circle end

Try it out

There are many possible answers. For example:

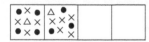 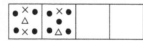

Matrices 2 (page 52)

Have a go

1.

2. a   Each column has same number of points on star or sides to
       shapes. Each column has one black star, one white shape on
       a black background and one white shape. Missing pattern is a
       white equilateral triangle on a black background.

---

b   Small black circle in lower right corner and rectangle across
    the top with vertical line shading – mirror image of patterns in
    the top row.
c   A white semi-oval with straight side across the bottom and
    diagonal stripes behind from bottom left to top right. White
    shapes in top row are reflected in the squares in the bottom
    row with same ones having a shaded background although the
    stripes stay the same rather than being reflected.

3. i   b   The top and bottom patterns combine to give the middle
           pattern in each of the other columns.
   ii d   Two of the patterns in each column combine to give the
           third pattern in each of the other columns.
4. The triangle will have grid lines, the top circle will be black and the
   bottom circle will have diagonal stripes from bottom left to top right.

Test yourself

1. c   **shading** – in each column the top cell (with its pattern of
       shaded squares) overlays the square in the middle row, and
       where shaded squares overlap they disappear from the bottom
       row. If they are white in both the top and middle row they
       remain white in the bottom row.
2. e   **shape** – (a) quarter circle in bottom-right corner of square,
       (b) small circles one each side of the short line; **shading** – (a)
       diagonal line shading from top left to bottom right in the quarter
       circle, (b) lower left small circle black, upper right circle white;
       position – short line from top-left corner to the centre of the box
3. c   **number** – two shapes (same as number of spots in adjacent
       square); **shape** – small rectangles (same as shape in adjacent
       square); **shading** – black (same as shading of corner triangle)
4. d   **number** – two circles as numbers in rows increase by one
       each time and numbers in columns reduce by two down the
       columns; **shading** – white circles (middle column all white)
       Distractors: **numbers** and **shading** of surrounding spots

Try it out

There many possible answers.

Maths workout 2 (page 54)

Rotating and translating images

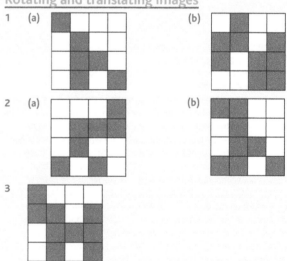

Working with numbers

1. (a) 8   (b) 11   (c) 11   (d) 22   (e) 19   (f) 12
   (g) 31   (h) 41   (i) 41   (j) 72   (k) 154   (l) 154
                                        (5 marks – 1 mark per row)
2. Each number is the sum of the two numbers above.
   (a) 30   (b) 10   (c) 20   (d) 20   (e) 25   (f) 25
                                        (4 marks – 1 mark per row)
3. Each number is half of the sum of the two numbers above.
   (a)   14 (+1, +2, +3, +4)
   (b)   24 (+2, +4, +6, +8)
   (c)   54 (+5, +10, +15, +20)   (3 marks)

Properties of 2D shapes

1. (a) pentagon   (b) trapezium   (c) equilateral triangle
2. 36°   3   48° (top) 60° (bottom)   4   51°   5   125°

Test yourself

The square on the left is folded in the way indicated by the arrows, and then holes are punched where shown on the third diagram. Identify the answer option that shows what the square would look like when it is unfolded. Circle the letter beneath the correct answer. For example:

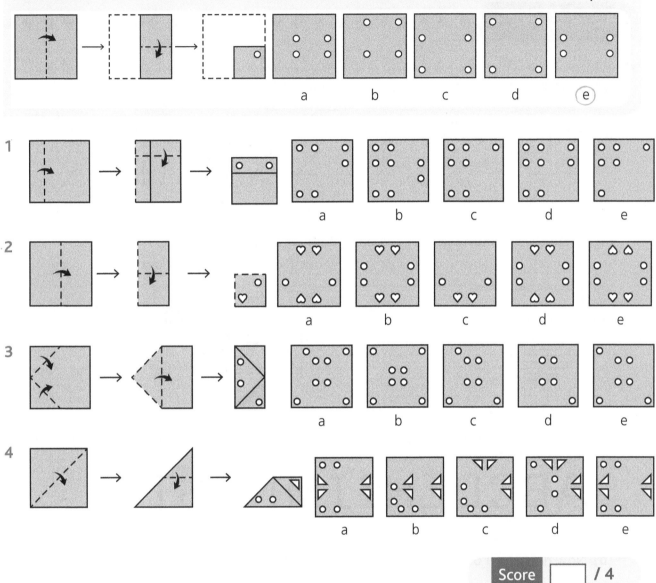

Score ☐ / 4

Try it out

Use dashed lines to show how to fold this sheet of paper so that punching just one circular hole will give this pattern when the paper is opened out.

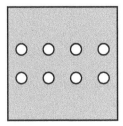

B Position and direction

Matching a single image 2

1 Which of these pictures has a vertical line of symmetry, with one half a reflection of the other half? Circle the letter and draw in the line of symmetry.

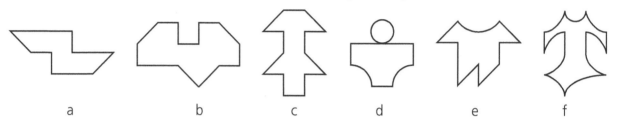

a　　　　b　　　　c　　　　d　　　　e　　　　f

2 Complete these pictures so that the dashed line is a line of reflection.

3 (a) Which of these letters have a vertical line of symmetry, where one half is a reflection of the other half? Write your answer on the line provided.

ABCDEHIJMNOST

_____

(b) Now draw each of the 13 capital letters rotated 90° clockwise.

Which ones now have a vertical line of reflection?

_____

Test yourself

The picture on the left is reflected in a vertical mirror line and is represented by one of the pictures on the right. Circle the letter beneath the correct answer. For example:

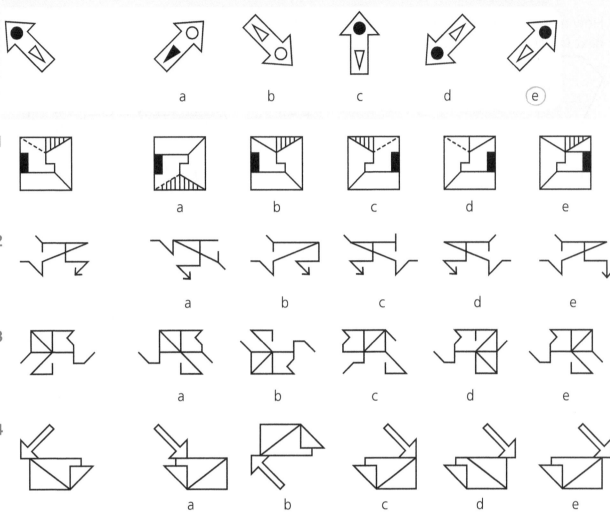

Try it out

Using at least four different colours, accurately fill in the squares and triangles in the picture below so that the dashed vertical line is a line of reflection.

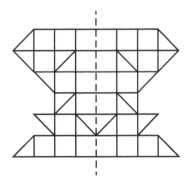

Translating and combining images 2

1   How many of each of these shapes are hidden in the picture on the right? Write the number next to each shape. Only count the shapes that are **not** rotated.

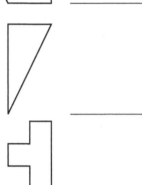

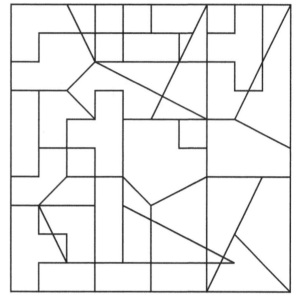

In the next two questions, the shape on the left is hidden in one of the pictures on the right. Find it and colour it in. The shapes are **not** rotated.

2

   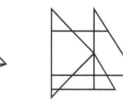   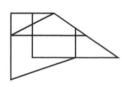   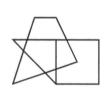

a   b   c   d

3

a   b   c   d

Marking or shading the shapes as you find them can be helpful.

Test yourself

The small shape on the left can be found in one of the pictures on the right. It might be made up of one or more pieces. Circle the letter beneath the correct answer. For example:

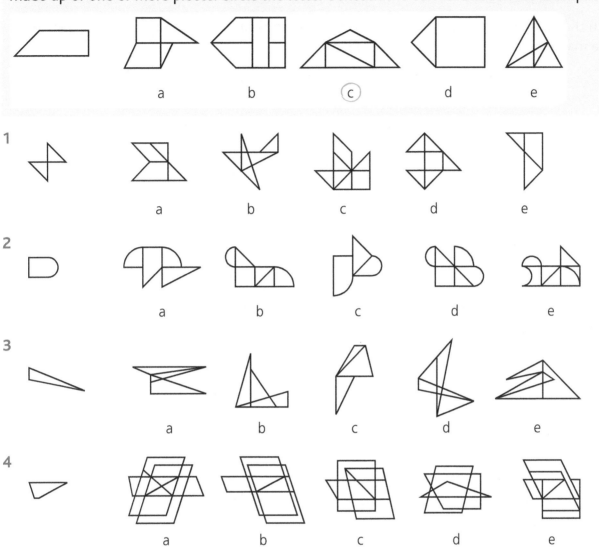

Try it out

Use a ruler to draw **eight** straight lines inside this rectangle so that the L-shape on the left is hidden within it. Ask a friend or parent to try to find the hidden L-shape.

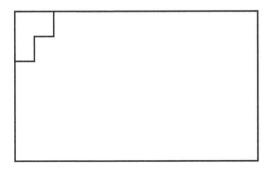

Score [ ] / 4

Matching a single image 3

Match the pictures in the first row to their identical copy in the second row. Write the letter of the matching picture in the space provided.

1

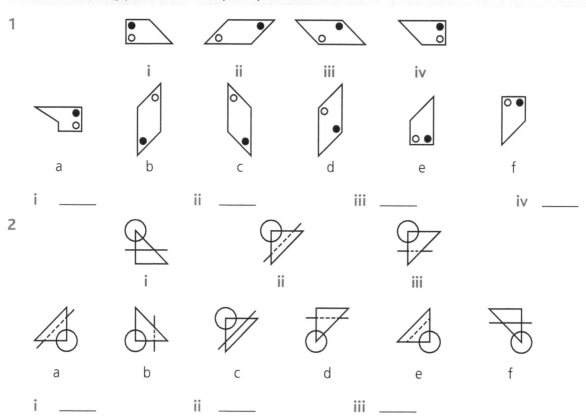

i _____   ii _____   iii _____   iv _____

2

i _____   ii _____   iii _____

In the next three questions, two of the three pictures in each are identical. Circle the letter of the **odd one out**.

3

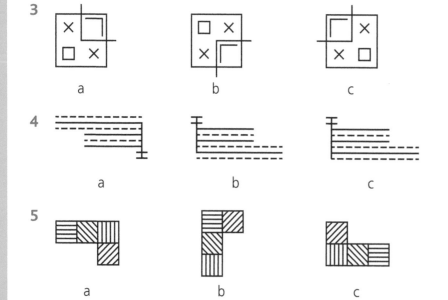

Test yourself

Look carefully at these pictures to identify the two that are identical. Circle the letters beneath the **two** identical pictures. For example:

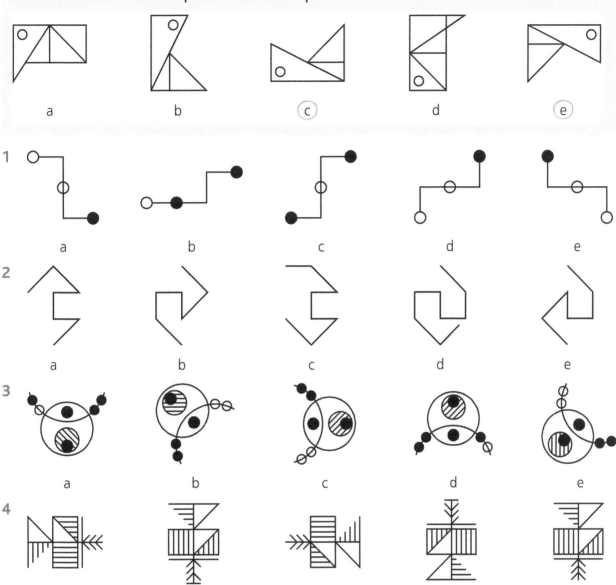

| a | b | (c) | d | (e) |

1
| a | b | c | d | e |

2
| a | b | c | d | e |

3
| a | b | c | d | e |

4
| a | b | c | d | e |

Score ☐ / 4

Try it out

Add patterns to these squares so that they are all quite similar but only **two** are identical. Try to include different **shapes**, different **shading styles** and a different **number** of elements. Ask a friend or parent to spot the identical patterns.

Translating and combining images 3

1   The square below has five shapes cut out of it. These shapes have been moved to the right
    of the square and have not been rotated. Find the five missing shapes and write their letters
    next to the correct number in the spaces provided.

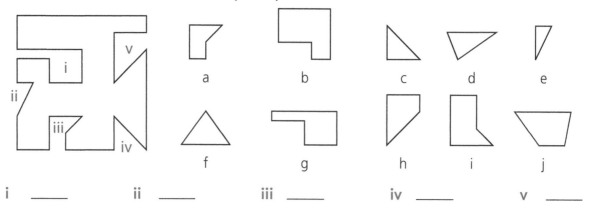

i _____   ii _____   iii _____   iv _____   v _____

2   Which two of the shapes in the second row can be put together to make each of the three
    shapes in the first row? Do not rotate them. Write the two letters of the shapes you select
    next to the correct numbers.

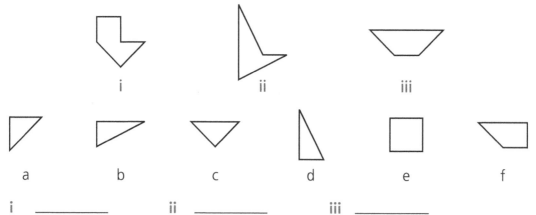

i _____   ii _____   iii _____

3   Which one of the three shapes on the right is **not** needed to make the shape on the left?
    Circle the letter beneath the correct answer.

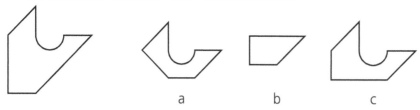

4   Which one of the four shapes on the right is **not** needed to make the shape on the left?
    Circle the letter beneath the correct answer.

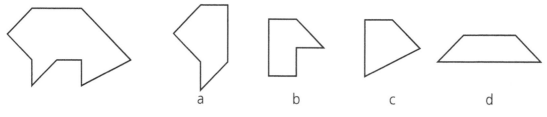

5   Which shape on the right has been cut out of the parallelogram on the left? The cut-out
    shape is **not** rotated. Circle the letter beneath the correct answer.

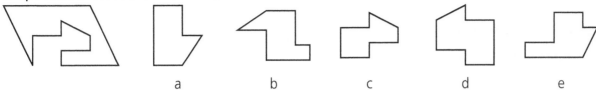

        a   b   c   d   e

Test yourself

When the smaller shape is removed from the larger shape a new shape is made. This new
shape is represented by one of the options on the right. Circle the letter beneath the correct
answer. For example:

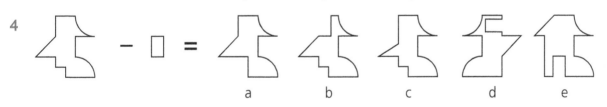

                        a   b   c   d   e

1

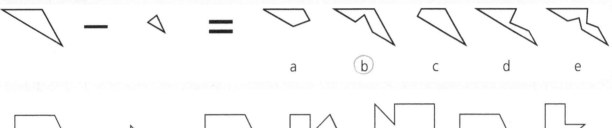

            a   b   c   d   e

2

            a   b   c   d   e

3

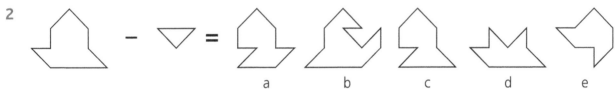

            a   b   c   d   e

4

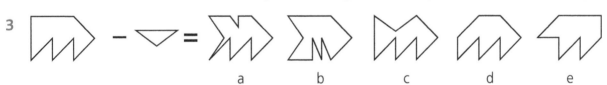

            a   b   c   d   e

Score [   ] / 4

Try it out

Try drawing your own question. Draw a small shape to be cut out of the large shape below.
Then provide five possible answer options. Only one of the answer options should match the
cut out.

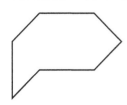

Translating and combining images 4

1   The pictures can be made using the individual blocks above them. Sometimes one block has
    been used more than once. Complete the table showing how many of each type of block will
    be needed to build the picture.

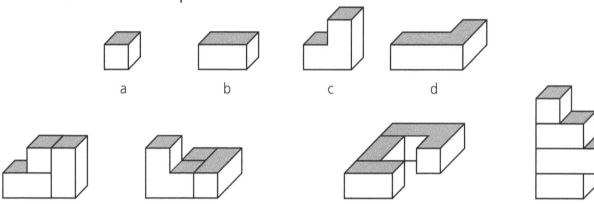

Picture	Block a	Block b	Block c	Block d
i				
ii				
iii				
iv				

Which pictures can be built with the blocks given? Circle the letter of your answer choice.

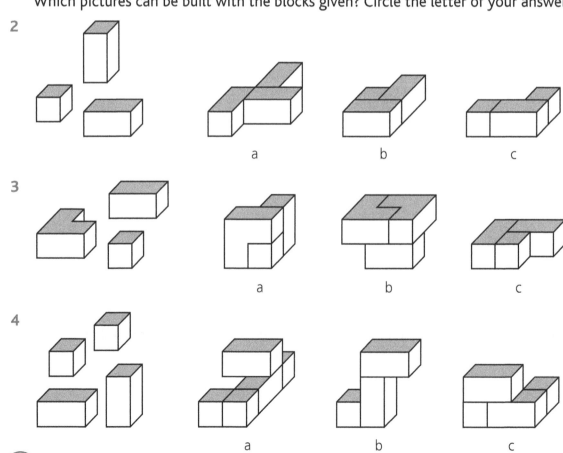

Test yourself

One group of separate blocks has been joined together to make the picture of blocks shown on the left of them. Some of the blocks may have been rotated. Circle the letter beneath the blocks that make up the picture. For example:

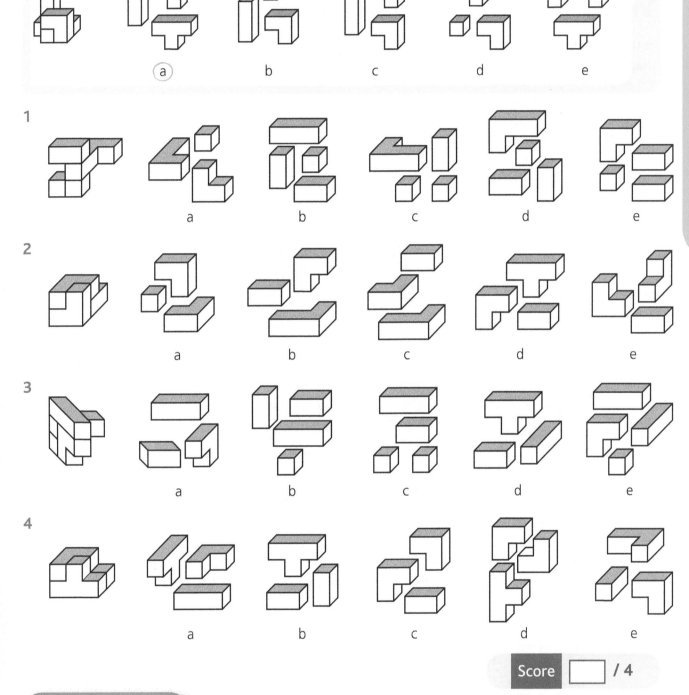

Try it out

How many different ways can you arrange these two blocks? At least one whole face must be touching the other block.

Maths workout 1

Many Non-Verbal Reasoning problems make use of mathematical skills and knowledge, so these pages contain some questions and puzzles to consolidate your mathematical skills, vocabulary and ideas. Keeping your maths skills sharp will help you to solve Non-Verbal Reasoning questions more quickly!

Working with 2D shapes

In the first two questions, which **four** pieces on the right can be put together to complete the 2D shape on the left? Circle the letters of your answer choices.

1

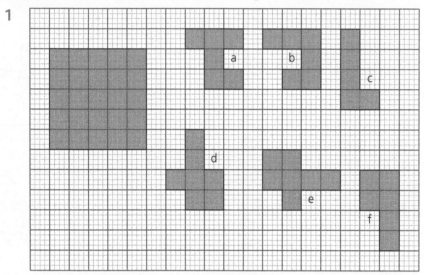

2

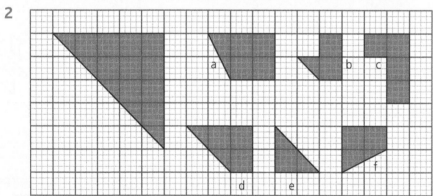

In the next two questions, move the 2D shapes by the number of squares instructed. Draw their new position and give the co-ordinates for the new position of point A.

3

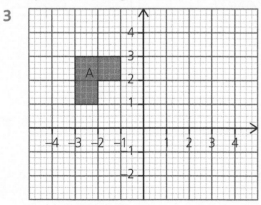

   i   right 4, down 2   new co-ordinates for point A (_____, _____)

   ii down 1, right 5   new co-ordinates for point A (_____, _____)

4

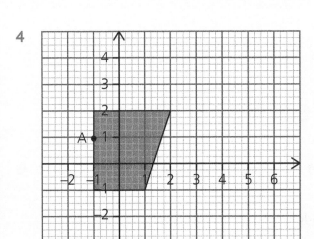

i   up 2, right 4   new co-ordinates for point A (_____, _____)
ii right 3, down 1   new co-ordinates for point A (_____, _____)

Score [   ] / 6

Working with 3D shapes

1   If each of these piles is made up of 1 cm cubes, what is the total surface area of each pile?

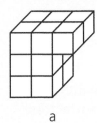   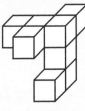   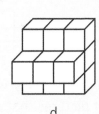   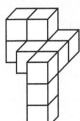

   a   b   c   d

Pile a: _____   Pile c: _____
Pile b: _____   Pile d: _____

2   If the outer surface of these piles of cubes were painted red, how many cubes would have:

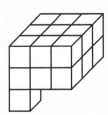

(a) i three painted faces? _____
    ii five painted faces? _____

(b) i three painted faces? _____
    ii five painted faces? _____   Score [   ] / 8

41

C Codes, sequences and matrices

Connections with codes 1

Have a go •

In the next two questions, work out the missing code letters and write them in the answer spaces provided beneath the pictures.

1

A ☐   BX   CZ   ☐ Y   CX
  a   b

2

LR   M ☐   ☐ R   LT   NS
        a   b

In the next two questions, write in the missing code letters. Then identify the features in each set of pictures that are not relevant to the answer and are included as distractors. Write these on the answer lines provided.

3

DL   EM   ☐ N   D ☐
                          a   b

Distractor 1: _____

Distractor 2: _____

4

AZ   C ☐   ☐ X   AY   BY
        a   b

Distractor 1: _____

Distractor 2: _____

Distractor 3: _____

5 Here there are three codes to work out.

    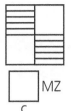

AMX BM BNZ CY MZ

a b c

Test yourself

Each letter represents an individual feature in the picture next to it. Work out which feature is represented by each letter. Apply the code to the picture in the box and circle the letter beneath the correct answer code. For example:

 SUW   TVX   TUY   SVZ

TVZ SUY SVX SUW TUZ

a b c d (e)

1

AG BF CG DF DG

a b c d e

AF BG CF

2

AQX BRZ BPY APZ ARY

a b c d e

APX BQZ BQY ARZ

3

BXG BYG AYH BXH AYG

a b c d e

AXG BYH AZH

4

NX MZ LY MX NZ

a b c d e

LX MY LZ NY

Score [ ] / 4

Try it out

Assign codes to the first four pictures so that the fifth picture has the code **BLT**.

(a) ____ (b) ____ (c) ____ (d) ____ BLT

Sequences 1

The first two questions contain a sequence of patterns with one pattern missing. Draw the missing pattern in the box provided.

1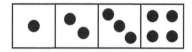

2

3 In the next sequence, one cross, one small circle and one small triangle are missing, each from a different box in the sequence. Work out the pattern and draw in the missing items.

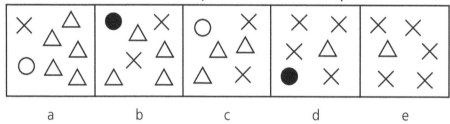

    a   b   c   d   e

4 Complete the sequence by adding shading to triangle **d**.

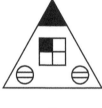   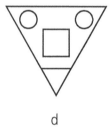

    a   b   c   d

5 The third picture in this sequence is incomplete. Work out what is missing and draw it in.

  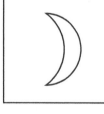

    a   b   c   d   e

Test yourself

The five boxes on the left show a pattern that is arranged in a sequence. Choose the answer option that completes the sequence when inserted in the blank box. Circle the letter beneath the correct answer. For example:

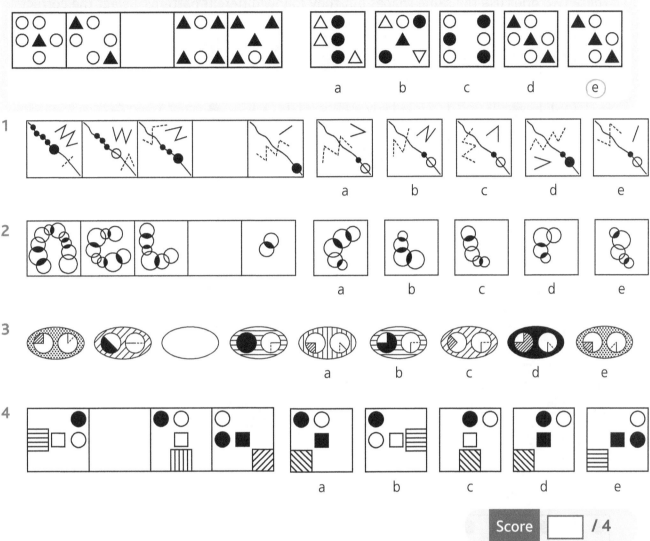

1

2

3

4

Score [   ] / 4

Try it out

Now create your own question. Use the pattern in the first box to make up a sequence of your own, leaving the fourth box empty. Then draw five answer options, with just one of them being the missing fourth pattern. Ask a friend or parent to identify the missing pattern.

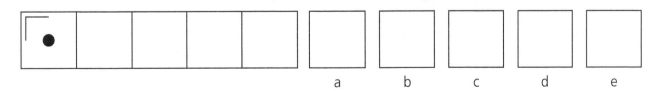

a   b   c   d   e

Matrices 1

1   These two grids use the same shapes but they follow different patterns. Select the correct shape for each empty square from the options given.

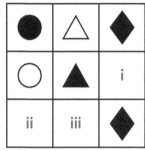   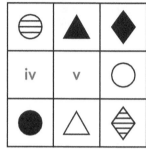   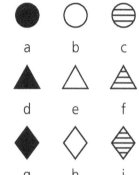

i ____   ii ____   iii ____   iv ____   v ____

Draw the missing pattern in the box provided to answer the next two questions.

2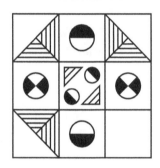

3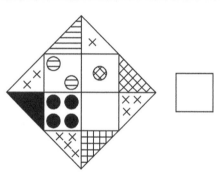

4   Identify the features that are changing along each row. Write them in the spaces provided.

(a) _____

(b) _____

(c) _____

(d) _____

5   The pattern in this shape is symmetrical. Draw dashed lines across the shape to show the lines of symmetry. Then draw the missing pattern in the box provided.

Test yourself

One of the options on the right completes the pattern in the grid on the left. Circle the letter beneath the correct answer. For example:

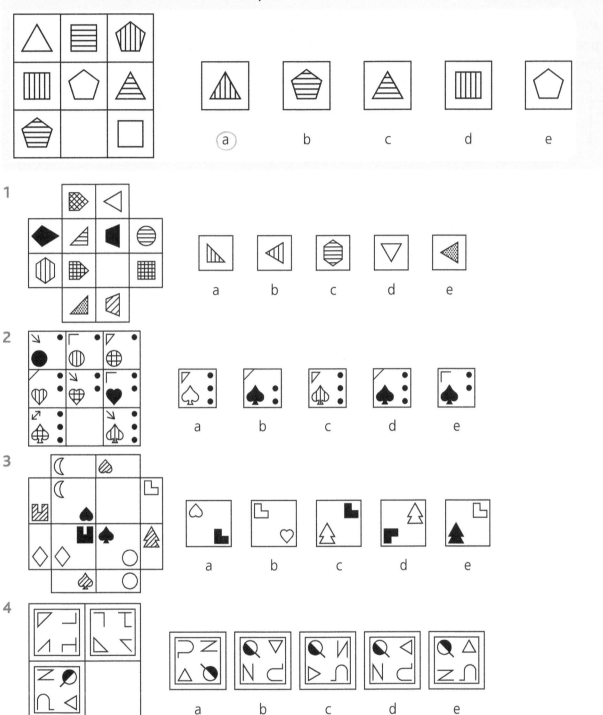

Score ☐ / 4

Try it out

Draw a 3 × 3 grid and create your own pattern using black circles and white circles with different shadings. Leave one square in the grid blank and give five possible answer options. Ask a friend or parent to identify the missing pattern.

Connections with codes 2

1   If the codes A, B and C represent a square, a circle and a triangle respectively, and L, M and N represent black, white and horizontal stripes, draw and shade the correct pattern in each box to match the codes given.

i   A [ ] N   ii B [ ] L   iii C [ ] M

iv B [ ] M   v   C [ ] N   vi A [ ] L

2   In these codes, the letter on the left relates to position and the letter on the right to number. Work out the code and fill in the missing letters.

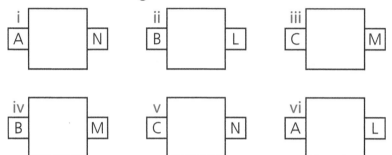

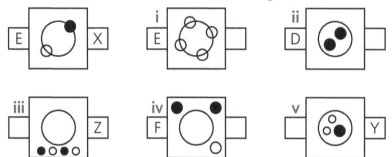

Codes have been assigned to the three pictures in each question. Work out the code and identify the feature represented by each letter. Write the features on the answer lines provided.

3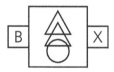

A _____
B _____
X _____
Y _____
Z _____

4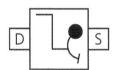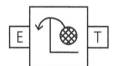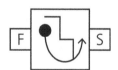

D _____
E _____
F _____
S _____
T _____

5   Complete the missing picture using the codes given.

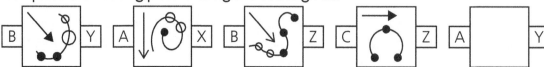

Test yourself

The letters in the boxes on the either side of these pictures each represent a feature of the pictures. Work out which feature is represented by each letter and apply the codes to the picture without any code. Circle the letter beneath the correct answer code. For example:

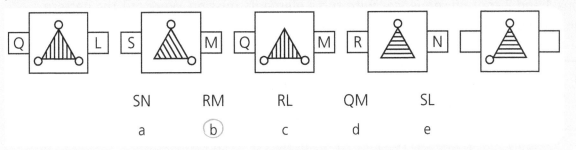

| SN | RM | RL | QM | SL |
| a | b | c | d | e |

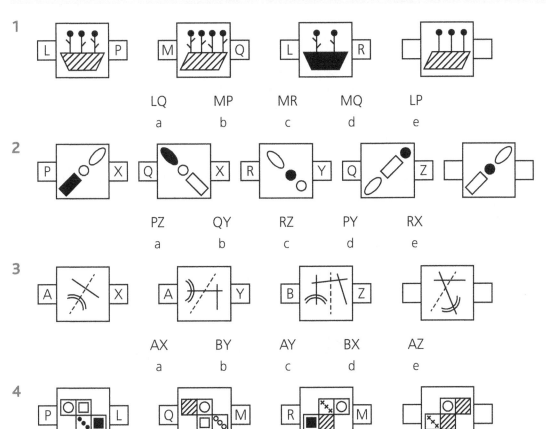

1

| LQ | MP | MR | MQ | LP |
| a | b | c | d | e |

2

| PZ | QY | RZ | PY | RX |
| a | b | c | d | e |

3

| AX | BY | AY | BX | AZ |
| a | b | c | d | e |

4

| PM | QL | QN | RL | RN |
| a | b | c | d | e |

Score ☐ / 4

Try it out

Make up your own codes question using the squares provided, giving five possible answer options.

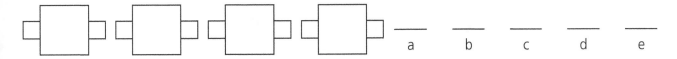

a b c d e

Sequences 2

Questions 1 and 2 contain sequences with three elements missing. Work out the pattern and then draw the missing elements.

1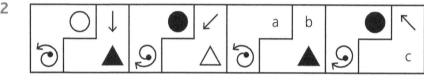

2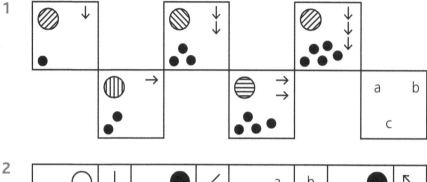

The next two questions contain a sequence with two alternating patterns. One pattern is missing from each sequence. Identify the missing pattern from the four options provided and circle the letter beneath it.

3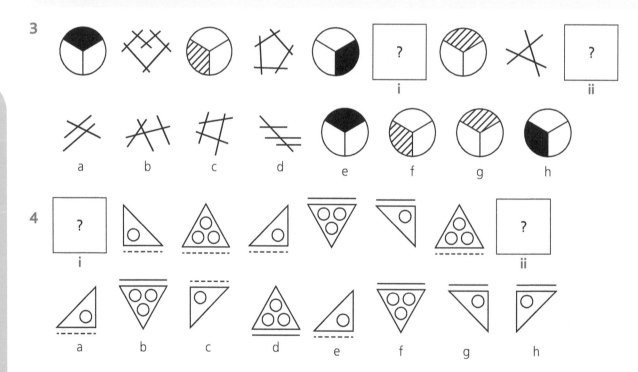

4

5 Continue the sequence by drawing the next three patterns in the sequence in the boxes provided.

      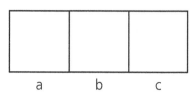

Test yourself

The five boxes on the left show a pattern that is arranged in a sequence. Choose the answer option that completes the sequence when inserted in the blank box. Circle the letter beneath the correct answer. For example:

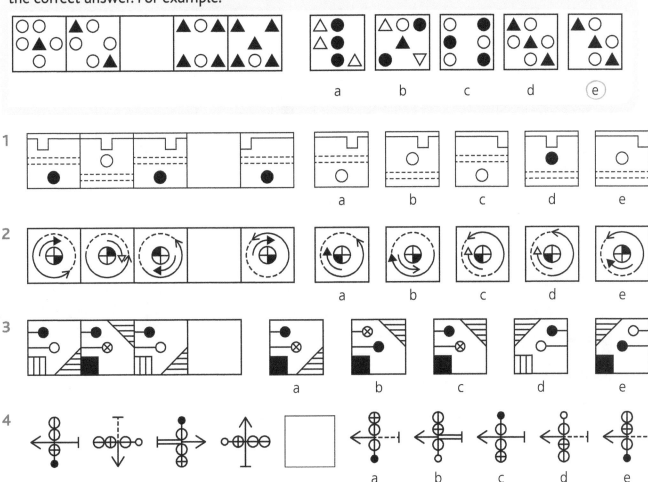

Try it out

Continue these two identical sequences in different ways. Draw the next two shapes in each sequence in the boxes provided.

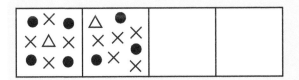

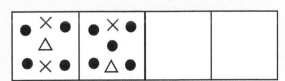

Score [   ] / 4

Matrices 2

Look at the patterns in the grids in the first two questions. Draw the missing patterns in the empty boxes.

1

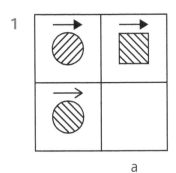

a

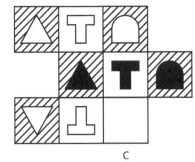

b

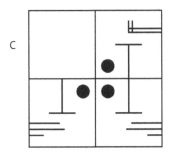

c

2

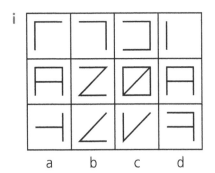

a

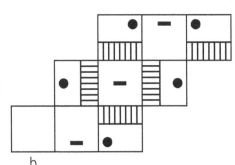

b

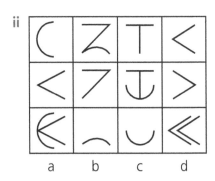

c

3 In the following grids, one column does not follow the pattern set out in the other columns. Which is the **odd** column?

4 Complete the next grid by adding the correct style of shading to the three shapes in the bottom right-hand square.

Test yourself

One of the options on the right completes the pattern in the grid on the left. Circle the letter beneath the correct answer. For example:

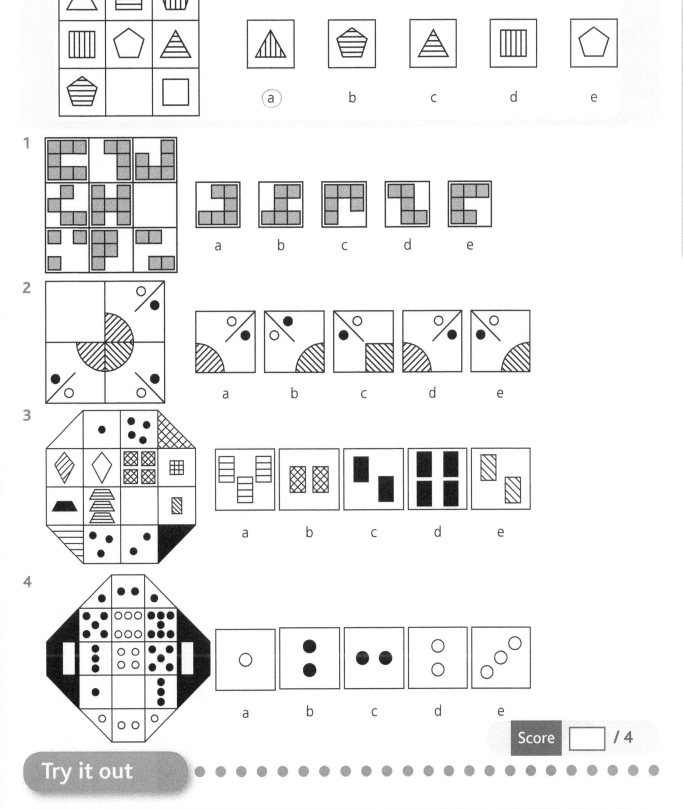

Score ☐ / 4

Try it out

On a separate piece of paper, draw a 4 × 4 grid and complete it with a shading pattern of your own. Leave one square empty and ask a friend or parent to identify the missing shading pattern.

Maths workout 2

Many Non-Verbal Reasoning problems make use of mathematical skills and knowledge, so these pages contain some questions and puzzles to consolidate your mathematical skills, vocabulary and ideas. Keeping your maths skills sharp will help you to solve Non-Verbal Reasoning questions more quickly!

Rotating and translating images

In the first two questions, the white squares in the grids are transparent. Rotate the grid as instructed. Imagine grid A is then placed over grid B. Draw the pattern you would see on grid C.

1 90° clockwise

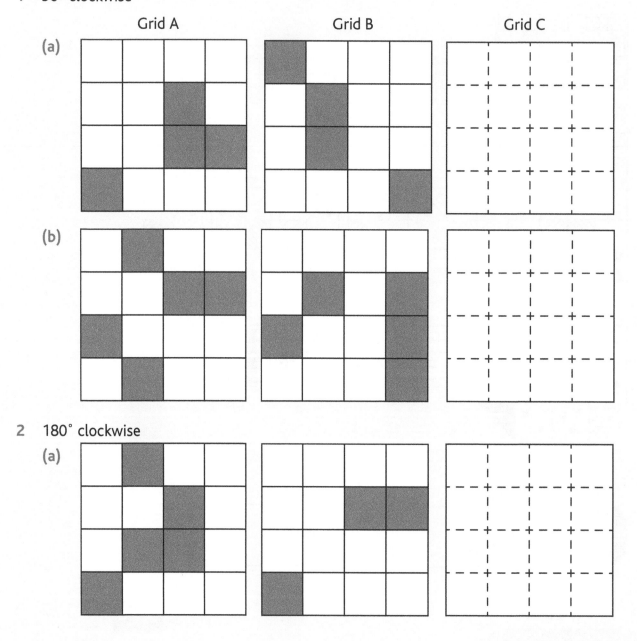

2 180° clockwise

	Grid A	Grid B	Grid C

(b)

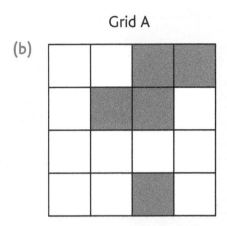 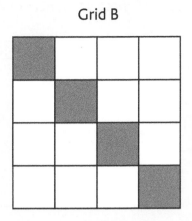 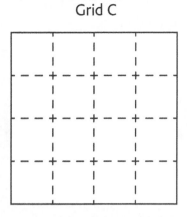

3 The white squares in these grids are transparent. Rotate grid A 90° clockwise and grid B 90° anticlockwise. Draw the pattern you would see when they are placed over each other on grid C.

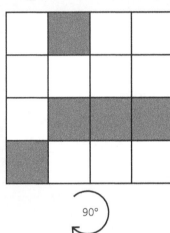 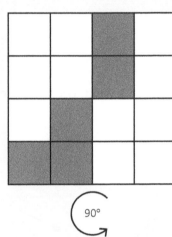 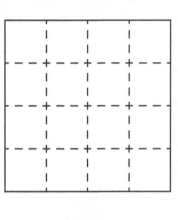

90°   90°

Score [ ] / 5

Working with numbers

In the following number patterns, work out how the numbers in each row have been calculated and complete the missing numbers.

1

1		3		5		6		5		3		1
	4		(a)____		(b)____		(c)____		8		4	
		12		19		(d)____		(e)____		(f)____		
			(g)____		(h)____		(i)____		31			
				72		82		(j)____				
					(k)____		(l)____					
						308						

2

5		15		45		15		5
	10		(a)____		30		(b)____	
		(c)____		30		(d)____		
			(e)____		25			
				(f)____				

55

3 What is the next number in each of these sequences? Write your answers on the lines provided.

(a) ☐4☐ ☐5☐ ☐7☐ ☐10☐ ____ (b) ☐4☐ ☐6☐ ☐10☐ ☐16☐ ____ (c) ☐4☐ ☐9☐ ☐19☐ ☐34☐ ____

Score ☐ **/ 12**

Properties of 2D shapes

1 What shape am I?

(a) I have five equal sides. _____

(b) I have four sides, but only one pair of opposite sides is parallel. I have no right-angles.

_____

(c) I have three sides and my three angles are all 60°. _____

The angles in a triangle all add up to 180°.

The angles around a point add up to 360°.

Vertically opposite angles at a point are equal.

Calculate the value of the missing angles and write them in these questions.

2

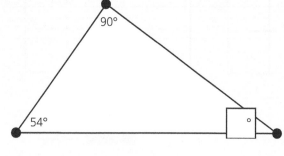

3

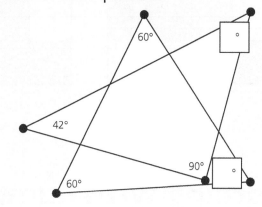

4   **5**

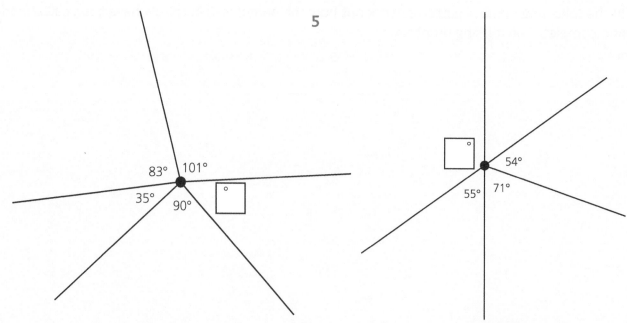

Score ☐ **/ 7**